AF813554

Redefining Sustainability

Jacinta Dsilva
Editor

Redefining Sustainability

International Conference on Net Zero Solutions and Circular Economy: Volume 1

Springer

Editor
Jacinta Dsilva
Global Sustainability Research Center
SEE Institute
The Sustainable City, Dubai, United Arab
Emirates

ISBN 978-3-031-88375-0 ISBN 978-3-031-88376-7 (eBook)
https://doi.org/10.1007/978-3-031-88376-7

© The Editor(s) (if applicable) and The Author(s), under exclusive license to Springer Nature Switzerland AG 2026

This work is subject to copyright. All rights are solely and exclusively licensed by the Publisher, whether the whole or part of the material is concerned, specifically the rights of translation, reprinting, reuse of illustrations, recitation, broadcasting, reproduction on microfilms or in any other physical way, and transmission or information storage and retrieval, electronic adaptation, computer software, or by similar or dissimilar methodology now known or hereafter developed.
The use of general descriptive names, registered names, trademarks, service marks, etc. in this publication does not imply, even in the absence of a specific statement, that such names are exempt from the relevant protective laws and regulations and therefore free for general use.
The publisher, the authors and the editors are safe to assume that the advice and information in this book are believed to be true and accurate at the date of publication. Neither the publisher nor the authors or the editors give a warranty, expressed or implied, with respect to the material contained herein or for any errors or omissions that may have been made. The publisher remains neutral with regard to jurisdictional claims in published maps and institutional affiliations.

This Springer imprint is published by the registered company Springer Nature Switzerland AG
The registered company address is: Gewerbestrasse 11, 6330 Cham, Switzerland

If disposing of this product, please recycle the paper.

Foreword

It was a pleasure for me to welcome all the participants and guests to SEE Global Sustainability Summit (SGSS) 2024. The seminal Summit focused on Net Zero Solutions and Circular Economy. The overwhelming response to our call-for-papers indicated the popularity of this Summit and confirmed that SGSS 2024 has become a world-wide forum to accelerate the transition toward a sustainable and resilient future. With a focus on advancing net zero strategies, circular economy principles, and sustainable innovation, this summit brought together experts, visionaries, and change-makers from across sectors and regions. The urgency of addressing climate change and resource scarcity demands new ways of thinking and collaborating. By fostering dialog on sustainable practices and promoting circular economy solutions, we aimed to empower industries and communities to reduce emissions, minimize waste, and maximize resource efficiency.

The Summit intended to involve various research areas, including Sustainable Built Environment, Climate Actions, Sustainability Leadership, Sustainability Education, Technology enabling sustainability and Circular Economy. As Chairman of SEE Institute and Summit President, I am confident that the insights shared in the book will drive transformative actions and inspire further innovation. Together, we can forge pathways to a sustainable, net zero world.

Thank you for your commitment and for being part of this vital journey.

Engr. Fares Saeed
Summit President and Chairman
SEE Institute
The Sustainable City, Dubai, United
Arab Emirates

Preface

We successfully curated and organized a two-day SEE Global Sustainability Summit on Net Zero Solutions and Circular Economy. The conference idea was well received by the team that it became broader and turned into a Sustainability Summit hosting academia, policy makers, students, startups, and corporates. The interesting part of the Summit was that research papers were not only presented by academicians but also by practitioners and industry professionals. Therefore, it is with great pleasure that I present '*Redefining Sustainability—Proceedings of the International Conference on Net Zero Solutions and Circular Economy*', a comprehensive two-volume work that brings together pioneering research, critical discussion, and innovative solutions from experts in the field of sustainability. This collection of chapters represents a significant contribution to the global dialog on sustainability, addressing key challenges and opportunities faced by both developed and developing economies in our collective pursuit of a highly sustainable future.

The idea behind the Summit sub-themes stem from the belief that our world urgently needs practical, actionable solutions to achieve sustainability. While recognizing the significant challenges associated with sustainable development is essential, focusing solely on those challenges can hinder progress. Instead, what the world needs is a platform for sharing and discussing solutions—both those that have been successfully implemented and those that have the potential to be. Our aim through both the volumes is to foster constructive discussion among the stakeholders and inspire innovation at all levels of the community. By emphasizing practical examples in the form of case studies and success stories, we intend to support UAE and other regions to adopt similar practices and scale up the impact of sustainability initiatives worldwide. This is only possible through collaboration, learning from one another and developing solution-oriented approach to turn sustainability from aspiration into a tangible reality.

The two volumes of the book aim to explore topics such as mitigating climate change and reducing our ecological footprint to fostering social equity and economic resilience, sustainability leadership to circular economy solutions encompassing a wide array of interconnected issues. The volumes further draw on diverse perspectives and present both theoretical insights and practical approaches to sustainable development, net zero strategies, and circular economy.

Volume 1: Redefining Sustainability—International Conference on Net Zero Solutions and Circular Economy

Sustainable Built Environment, Rating Systems, Climate Actions, and Resilient Cities

The first volume focuses on climate actions and net zero solutions largely in the built environment due to its impact on the planet. As cities continue to grow, there is an increasing need to ensure that urban development minimizes environmental impact while promoting social well-being. This volume covers several key topics, including climate actions, sustainability rating systems for buildings, and strategies for creating resilient cities and communities. The chapters under this volume highlight different solutions pertaining to the role of geotechnical engineering, adoption of renewable energy in the UAE, water availability and high temperature impact on the biodiversity, micro-air quality and green infrastructure. Furthermore, under the theme of rating systems in a sustainable environment, the chapters focused on developing new framework and standards for evaluating social and economic dimensions, standards for examining consumer adoption of sustainable brands, as well as sustainability maturity assessment.

Under the theme resilient cities 2050 and sustainable environment, the discussions on resilient cities are particularly timely, given the increasing frequency and intensity of natural disasters linked to climate change. the authors focused on chapters such as evaluating visual comfort through dynamic day lighting spaces, all-consequences calculator for calculating roadworks, evaluating CO_2 emissions of an industrial warehouse, influence of heritage design and sustainable interior environment, Aquaponics for energy saving and food production, and assessing sustainable materials.

Overall, the contributors to this volume address the challenges of balancing economic growth with environmental conservation. They delve into issues such as reducing greenhouse gas emissions in the construction sector, developing eco-friendly infrastructure, and promoting energy efficiency through cutting-edge technologies. The research presented highlights the critical role that cities play in achieving global climate goals, while offering practical recommendations for policymakers, urban planners, and industry leaders.

The ultimate aim of volume I is to assist readers in how urban areas can prepare for and recover from such events, emphasizing the importance of infrastructure resilience, community engagement, and long-term planning. "Volume 2: Redefining Sustainability - Proceedings of the International Conference on Net Zero Solutions and Circular Economy Sustainable Leadership, Sustainability Education, Circular Economy, and Technology as an Enabler". The second volume shifts the focus to the broader aspects of sustainability, including sustainable leadership, sustainability education, the circular economy, and the role of technology in enabling sustainable practices. This volume builds on the idea that sustainability is not limited to environmental considerations; it also encompasses social and economic dimensions. In the section on sustainable leadership, the authors discuss how businesses and organizations can take a proactive role in driving sustainability. The chapters focus on several aspects such as developing human capital to drive social impact, integrating global leadership competencies for transformative leadership, human capital development and sustainable careers, and building a global sustainable future. The chapters in this section explore leadership models that prioritize not just profit, but also people and the planet. Case studies from diverse sectors and countries illustrate how forward-thinking leaders can foster a culture of sustainability within their organizations and across supplychains. The volume further goes into Sustainability education since if we need change in human behaviour we need to make major reforms in the way we teach particularly sustainability related discussions. Therefore, the chapters in this section highlight the topics such as greening he learning experience in higher education, ranking higher education institutions on their sustainability efforts, integrating higher education institutions to bridge the future workforce gap, eco-education, best practices in fostering green skills and designing a youth environmental stewardship training program. The circular economy (CE) is another key theme in this volume, offering insights into how businesses can shift from traditional linear models of production and consumption to more sustainable, regenerative practices. The contributors examine innovative strategies for reducing waste, reusing materials, and creating closed-loop systems that benefit both the environment and the economy. The chapters in this section highlight technology and promotion of circular economy and global sustainable growth, developing brand legitimacy through circular economy, close the PET plastic recycling loop, the growth of circular fashion economy, and CE practices for achieving SDGs. Finally, the role of technology as an enabler of sustainability is explored in depth. This section highlights how advancements in digital technologies, such as artificial intelligence (AI), the Internet of Things (IoT), and blockchain, are transforming industries and driving sustainable development. From improving resource efficiency to enhancing transparency in supply chains, technology has the potential to revolutionize our approach to sustainability. The chapters focus on blockchain trust model, sharding as a sustainability blockchain optimization tool, integrating NetZero IOT, Cloud, and AI for Smart Energy Management, Sustainable Villa Retrofitting Technologies to Achieve Net Zero, role of digital marketing, and Smart university campus living lab for climate action. **Bridging Theory and Practice**. What makes this collection particularly valuable is its ability to bridge the gap between theory and practice. The papers presented in these volumes offer both deep theoretical insights and practical solutions that can be implemented by practitioners in the field. Whether it is through the adoption of green technologies, sustainability leadership and education, the development of sustainable policies and frameworks, or the promotion of community-based climate resilience strategies, the research showcased here provides a roadmap for achieving a sustainable future.

Dr. Jacinta Dsilva
Director of Research and Global Partnership
SEE Institute
The Sustainable City, Dubai, United Arab Emirates

Acknowledgments

I would like to extend my deepest gratitude to the authors, and reviewers, of the SEE Global Sustainability Summit 2024 for their invaluable contribution to this publication. Their dedication to advancing knowledge and fostering collaboration in the field of sustainability is evident throughout this book. The editor would also like to thank Lydia Cherian, Myriam Larbi and Sahil Buch for your continuous support in finalizing the book. We thank the team at Springer Nature for their continued support in disseminating research that makes a meaningful impact on global challenges.

I hope that the insights and innovations presented in these volumes will inspire further research, encourage cross-disciplinary collaboration, and contribute to the development of sustainable solutions that will benefit both present and future generations.

About SEE Institute

SEE Institute is a hub for research and education in the areas of sustainability, seeking to empower the growth of sustainable practices and transition to a net zero world. The region's first education provider fully dedicated to sustainability education. The institute is committed to support the demand for skilled sustainability experts. At SEE Institute, our professional education programs are designed to prepare professionals for the ever evolving filed of sustainability in the built environment sector. Our Sustainability Research Center plays a pivotal role in advancing sustainable practices through research initiatives conducted in collaboration with industry, governments, and academia. Our emphasis is upon projects that focus on the three pillars of sustainability, sustainability in the built environment, sustainability leadership, and net zero strategies. Our outcomes are shared through peer reviewed journal articles, white papers and recommendation reprts as well as conducting impact assessment studies. In addition, through our Incubation program we support and invest in startups that offer innovative sustainability solutions. Our Incubation Center supports startups in growth, scaling and validating their ideas into reality. Overall, SEE Institute's aim is to enable growth through education, research and innovation to speed up the implementation of sustainable solutions to achieve national and international sustainability goals.

Contents

About the Editor

Dr. Jacinta Dsilva is a renowned sustainability researcher and strategist with over 20 years of academic and industry experience. She holds the position of Director of Research, Living Lab and International Partnerships at SEE Institute. Her expertise lies in the intersection of sustainability and consumer behavior, and she has worked at prestigious universities in the UAE, India, and China. With a strong focus on research, Dr. Dsilva has published numerous books, book chapters, and academic papers on topics such as Green Technologies, Industry 4.0, Strategic Corporate Governance and Sustainability.

She has led research teams, established collaborative partnerships with high-level stakeholders, and developed frameworks for conducting research in circular economy and climate change. Driven by her commitment to sustainability, she has established a research center and living lab to facilitate impactful research, validation of new sustainability ideas and knowledge dissemination. Dr. Dsilva's strategic insights and market analysis skills have supported businesses in reducing plastic waste and developing sustainable products and services. With her leadership and expertise, she continues to make significant contributions to academia, industry in the pursuit of a sustainable future.

Recognized for her thought leadership, Dr. Dsilva has published numerous academic papers in reputable Scopus-based, UGC-Care and ABDC journals as well as presented her research findings at prestigious local and international conferences. She has advised universities and schools on embedding sustainability in their curricula, demonstrating her commitment to shaping future leaders with a sustainability mindset. Recently, she led and organized SEE Global Sustainability Summit 2024 attended by over 500 academics, startups, industry and government officials across the world. She has also worked with National Dialogues organized by the UAE Ministry of Climate Change and Environment and played a significant role in organizing crucial discussions during COP28 alongside the Ministry of Education.

Part I
Climate Actions and Net Zero Solutions

Chapter 1
Climate Action and Net Zero: The Role of Underground Space and Geotechnical Engineering

Dexter Vernon Lloyd Hunt, Farrukh Hafeez, and Ian Jefferson

Abstract Assessment and Quantification of the carbon impact of Geotechnical Assets and Underground Construction are at an early stage of development. However, the road to achieving Net Zero is very much upon us and a path on which we all must quickly travel. Unfortunately, a lack of guidance on the suitability of existing carbon accounting tools is apparent in many sectors, not least geotechnical Engineering, moreover the role that Underground Space and Geotechnical Engineering can play in net zero is very much understated. This paper seeks to open-up discussion in this area, helping bridge a gap in knowledge to help practitioners and engineers make informed choices when it comes to carbon savings, not least in application to new (and existing) Geotechnical Applications and Underground Space.

Keywords Net zero · Geotechnical engineering · Underground space · Carbon savings · Sustainability in construction

Introduction

In 2015, the historic Paris Agreement to reduce global warming and build resilience to climate change was adopted by 196 countries. Its overall aim was to *'Limit Global Warming to no more than 1.5 °C'* (UNFCC, 2023). In the last two years, science has affirmed an ever-diminishing window of opportunity to make a positive change and

D. V. L. Hunt · F. Hafeez · I. Jefferson (✉)
Department of Civil Engineering, School of Engineering, University of Birmingham, Edgbaston, Birmingham, UK
e-mail: i.jefferson@bham.ac.uk

D. V. L. Hunt
e-mail: d.hunt@bham.ac.uk

F. Hafeez
e-mail: f_hafeez@yahoo.com

F. Hafeez
University of Birmingham, Dubai Campus, Dubai, United Arab Emirates

© The Author(s), under exclusive license to Springer Nature Switzerland AG 2026
J. Dsilva (ed.), *Redefining Sustainability*, https://doi.org/10.1007/978-3-031-88376-7_1

strengthen climate action. As such post COP 26, 27, and 28 climate talks have led to countries revising Nationally Determined Contributions (NDCs) previously agreed in Paris (UNFCC, 2023). As such it is currently estimated that by 2030, countries around the globe must cut their carbon emissions by at least 45% compared to 2010 levels to keep global warming to 1.5 °C. Whilst hosting COP 28, the UAE announced a landmark Global Decarbonisation Accelerator, to fast-track a new, global energy system. Infrastructure per se is estimated to be responsible for around 70% of carbon emissions globally, hence ensuring an imperative whereby the provision of low carbon infrastructure is the norm is key to transitioning to net-zero emissions, which must be fully complete by 2050. More broadly, Geotechnical Engineering and the role played by use of underground space and infrastructure cannot be underplayed or overlooked when it comes to achievement of SDGs (e.g. Clean Water, Clean Energy, Responsible Consumption, and Climate Action), Net Zero and other related 2030 and 2050 targets.

Life-Cycle Assessment of Geotechnical and Underground Assets

'LCA is used to investigate the environmental impact of a defined system through its life cycle'. (EEA, 2024). In Geotechnical construction, the system could be a material manufacturing process (e.g. cement), the fabrication of a construction component (e.g. precast pile, geogrid, ground anchors), a geotechnical construction element (e.g. an earth reinforced wall, stabilized slope, vertical drains setup) or even an entire underground construction project (e.g. tunnelling or street works). In respect to Geotechnical Engineering and more broadly the use of underground space, it relates to the process by which Geotechnical assets are created to the point at which they are managed and their end-of-life. Therein the Life Cycle for geotechnical Assets consists of 4 key stages (i.e. A–D, shown in Fig. 1.1).

The Cradle-to-Gate component (A1–A3) includes all key materials required to construct a Geotechnical Asset. These aspects include, but are not limited to, extraction and transportation and production of resources for producing construction materials (e.g. Rebar, Aggregates, Cement, Bitumen, Tarmac, Asphalt and the like). As part of the construction process (A4–A5), these component materials are transported to site ready for installation. Geotechnical procedures and operations, depending on the scale and type have numerous machine options, not least Piling rigs, Excavators, Backhoes, Trucks, Rollers, Generators, TBMs …to name but a few. In-use operation of geotechnical assets takes us from completion of the geotechnical asset to end-of life, thus includes maintenance, Repair, Replacement and Refurbishment or upgrading (i.e. B2–B5).

B1 refers to *'use'* and, depending on where the boundary is drawn, is less relevant to the geotechnical asset—in other words, some geotechnical assets (e.g. a piled foundation an earth retained wall) do not have a carbon associated with their actual

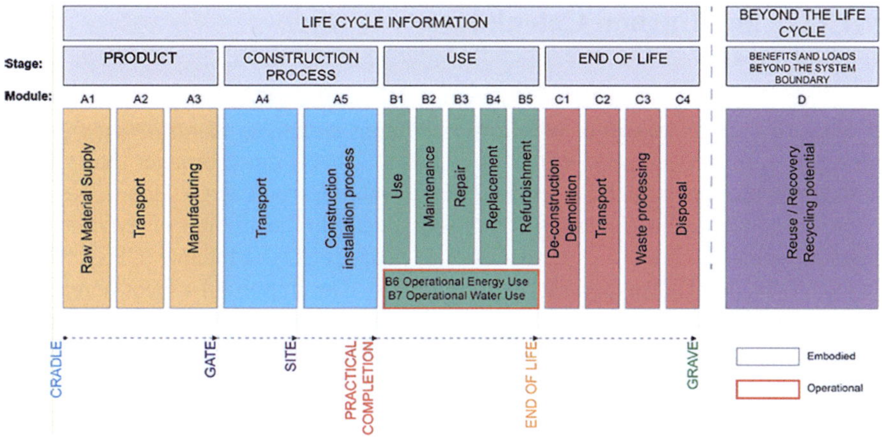

Fig. 1.1 Life cycle stages (Extract from BS EN 15978/PAS2080)

use, whereas others (e.g. a road tunnel) might in terms of lighting and ventilation. Moreover, some innovative geotechnical assets such as geothermal piles, could have a net positive impact in terms of carbon savings from earth extracted energy (Sadeghi & Singh, 2023) Additionally, some carbon emissions can be indirectly related to the use of a geotechnical asset (e.g. traffic emissions on a road/rail structure). The same can be said about operational Energy use (B6) and Water use (B7), which form a necessary integral part of achieving B1–B5.

Aspects related to End of Life (C1–C4) and beyond grave (D) allow for movement towards a more circular economy type approach to geotechnical assets. A full life cycle assessment employs a series of stages to conduct a rigorous LCA assessment. Whilst the focus is currently on achieving Net Zero (with the category being Greenhouse Gases and the normalized measure being CO_{2e}). However, it should not be forgotten that some 17 Environmental Categories exist (e.g. Human Toxicity, Water Extraction, Acid Deposition, Ozone Depletion, Eutrophication to name a few). Therefore, focusing on Net Zero should not be to the detriment of these other categories. Moreover, sustainability should also seek to balance impact(s) across and between all three pillars, asking pinch-point questions such as: is it viable, is it equitable and is it bearable? Moreover, engineers should ask whether the geotechnical design and use of underground space is resilient to a future which may be different to what we know now (Hunt et al, 2012a, 2012b).

In fact, this should be considered as the starting point for the construction of any Geotechnical Asset and for the true cost assessment of any underground construction project. Whatever is being considered some sort of assessment needs to take place, which requires Indicators, Metrics, Benchmarks and Assessment Frameworks or Tools. As the current focus is on Net Zero, it is worth taking time to see which carbon calculator tools are currently available and how they contrast and compare when it comes to the ability to deliver whole life carbon reporting for underground space and geotechnical assets (See Section "Net Zero and Carbon Calculators").

Net Zero and Carbon Calculators

Net zero, according to the UN this means: *'Cutting greenhouse gas emissions to **as close to zero as possible**, with any remaining emissions re-absorbed from the atmosphere, by oceans and forests.'* (UN, 2023). This is much broader than Carbon neutrality which specifically refers to the removal of Carbon alone. Hence, Net Zero in respect of Geotechnical Assets refers specifically to the amount of removal of all greenhouse gases (GHGs)—such as Carbon Dioxide (CO_2), Methane (CH_4), Nitrous Oxide (N_2O) and fluorinated Gases (UN, 2023). The important aspect here is that Net Zero requires us to take a much broader view than just *'in-use'* emissions. In other words, embracing more readily the life cycle approach (LCA) of a product or process (See Section "Life-Cycle Assessment of Geotechnical and Underground Assets"). UAE is very proactive and spearheading the climate agenda is certain areas for example UAE Net Zero by 2050 strategic initiative and Al Sa'fat–Dubai Green Building System (Government of Dubai, 2023), however standardized net-zero calculators are rare in this region particularly specific to underground construction. Hence opening discussion in this area is very relevant to the UAE and wider region because UAE aims to make Emirates the first Middle EAST and North Africa (MENA) nation to achieve net-zero emissions by 2050 (UAE Government Portal). Recent infrastructure developments across the UAE also highlights this need. For example, inauguration of two double tracks with five tunnels in Sharjah in 2019, which are more than six kilometres of the total length of the road (Khormun, 2018). Table 2 (Appendix) gives a summary of some of the available carbon assessment tools. Whilst the list is not exhaustive, it highlights the need to consider what type of tool it is (i.e. web-based vs excel-based vs software instal), what scope the tool has (including region for application), what part(s) of an LCA it considers (A–D), what database(s)/data it uses and how up-to-date these are. Considering the later point, a centralized repository or portal, such as that for Environmental Product Declarations (EPDs) (EPD, No Date), is required in addition to portals for databases (i.e. for Carbon and Energy) that house free-to-access inventories. One such portal is the Built Environment Carbon Database (BECD), designed to become the main source of carbon estimating and benchmarking for the construction industry (BECD, 2023). Unfortunately, whilst no individual tool can collectively assess the carbon cost of underground space—individual tools have been developed for certain aspects, such as Infrastructure, Roads/highways, Rail, Trenchless Technology and Geotechnical projects (with a focus on foundations). Whilst numerous tools assess cradle-to-gate (i.e. A1–A3), very few assess all four stages (i.e. A through to D). Hence much work, such as development of bespoke tools, is required to meet these requirements. In addition, carbon tools should be cross compared to ensure consistency, transparency and 'fair play' exist.

Challenges of Underground Space Use in a Net-Zero World

In October 2022, whilst presenting at the GE Basements and Underground Structures conference Tim Chapman, who is responsible for net zero carbon infrastructure at Arup, posed the following question to the construction industry: *Creating underground space involves huge carbon expenditure—how can this be justified?* Whilst this may be true, there are ways to reduce its impact and ways to optimize the use of Underground space to facilitate the Net-Zero journey. For example, the International Energy Agency (IEA) analysis of net-zero pathways highlighted the necessity for CO_2 capture and underground storage (CCUS), suggesting an ambition to grow from around 40 Mt/year in 2020 to more than 5000 Mt/year by 2050 to support Net-Zero ambitions (IEA, 2020). Therein, the IEA sees a future with Carbon management services—transporting and storing CO_2 in large quantities—a global industry with new underground infrastructure supporting emissions reduction (IEA, 2020). Underground space can also be used for storage of other more hazardous substances such as radioactive waste (Hunt et al., 2012a) to support Nuclear and Net Zero. Additionally, underground constructions/dwellings have a proven track record in being more energy-efficient (due to insulating properties of the ground) compared to traditional buildings (Xie et al., (2022) and organizing utility infrastructure in multi-utility tunnels (MUTs) has been shown to have a whole host of sustainability, and cost benefits (including, but not limited to Lower maintenance, reduced disruption and increased safety) in addition to carbon and resilience advantages (Hunt et al., 2012a, 2012b; Hojatti et al., 2023). The low carbon effects of utilizing urban underground space are increasingly being reported (Qiao et al., 2019). However, a quantitative analysis of this is still necessary to encourage the wider benefits of its adoption (Xia et al., 2024).

Conclusions

This paper has opened-up discussions in the areas of Sustainability, Climate Action and Net Zero underlining the role Underground Space and Geotechnical Engineering can play. This will help bridge a gap in knowledge to help practitioners and engineers make informed choices when it comes to carbon savings. Combined with an improvement in tools it is necessary to upskill employees to help them become champions in the understanding of sustainability, net zero, life cycle assessment (LCA) and circular economy (CE) within their organizations. These can be facilitated through innovative bootcamps, which have proved to be highly successful for those within and outside the engineering community (Johal, 2023).

References for Tools

Asphalt Pavement Embodied Carbon Tool (asPECT). https://trl.co.uk/Uploads/TRL/Documents/asPECT%20Protocol%20v4.2%20%20-%20August%202020%20-%20clean.pdf
Built Environment Carbon Database (BECD). https://www.becd.co.uk/
Carbon Trust. https://www.carbontrust.com/resources/sme-carbon-footprint-calculator
European Federation of Foundation Contractors and Deep Foundation Institute Carbon Tool (EFFC DFI). https://www.effc.org/how-we-operate/eco%e2%82%82-foundations/
eToolLCD. https://etoolglobal.com/about-etoollcd/
Environment Agency (ERIC) Tool. https://assets.publishing.service.gov.uk/government/uploads/system/uploads/attachment_data/file/571707/LIT_7067.pdf
Future Highways Research Group (FHRG) Future Highways Research Group | ADEPT (adeptnet.org.uk).
GaBi Software. https://gabi.sphera.com/america/index/
Interoperable Carbon Assessment Toolkit (iCAT) GtR (ukri.org).
IStructE Guide and Structural Carbon Tool (SC). https://www.istructe.org/resources/guidance/the-structural-carbon-tool/
National Highways (NH) Tool. https://nationalhighways.co.uk/suppliers/design-standards-and-specifications/carbon-emissions-calculation-tool/
One Click LCA World's fastest Building Life Cycle Assessment software—One Click LCA.
Rail Carbon Tool (RSSB). https://www.rssb.co.uk/en/sustainability/rail-carbon-tool
Trenchless Technology Calculator (TTC). https://nastt-bc.org/carbon-calculator/
Sima Pro. https://simapro.com/

References

BECD. (2023). The Built Environment Carbon Database (BECD). https://www.becd.co.uk/
Civic facilities | The Official Portal of the UAE Government. (2021). Retrieved August 2, 2024, from U.ae website: https://u.ae/en/information-and-services/infrastructure/civic-facilities
EEA. (2024). Life cycle assessment. https://www.eea.europa.eu/help/glossary/eea-glossary/life-cycle-assessment#:~:text=Life%2Dcycle%20assessment%20(LCA),use%20efficiency%20and%20decreasing%20liabilities
EPD International. (2024). Retrieved August 2, 2024, from Environdec.com website. https://www.environdec.com/home
Government of Dubai. (2023). *Al Sa'fat—Dubai Green Building System* (2nd ed.)
Hammond, G., & Jones, C. (2011). Inventory of Carbon and Energy Database Embodied Carbon—The Inventory of Carbon and Energy (ICE) (greenbuildingencyclopaedia.uk).
Hojjati, A., Hunt, L., & Rogers, F. (2022). Assessing the Carbon Cost of Utility Installation via Multi-Carbon Management in Buildings and Infrastructure. (n.d.). Retrieved from https://www.bsigroup.com/siteassets/pdf/en/insights-and-media/insights/brochures/pas_2080.pdf
Hojjati, A., Hunt, D. V. L., & Rogers, C. D. F. (2023). Assessing the carbon cost of utility installation via multi-utility tunnels (MUTs). In *Proceedings of the Canadian Society of Civil Engineering Annual Conference 2021*. Springer Nature Singapore.
Hunt, D. V. L., Jefferson, I., & Rogers, C. D. F. (2012a). Testing the Resilience of Underground Infrastructure Solutions through an Urban Futures Methodology. *Proceedings of REAL CORP, 2012*, 825–834.
Hunt, D. V. L., Nash, D., & Rogers, C. D. F. (2014). Sustainable utility placement via multi-utility tunnels. *Tunnelling and Underground Space Technology, 39*, 15–26. https://doi.org/10.1016/j.tust.2012.02.001

Hunt, D. V. L., Lombardi, D. R., Atkinson, S., Barber, A. R. G., Barnes, M., Boyko, C. T., Brown, J., Bryson, J., Butler, D., Caputo, S., et al. (2012b). Scenario archetypes: Converging rather than diverging themes. *Sustainability, 4*, 740–772. https://doi.org/10.3390/su4040740

IEA. (2020). CCUS in Clean Energy Transitions—Analysis. https://www.iea.org/reports/ccus-in-clean-energy-transitions

Johal, K. (2023). Net Zero Together Bootcamp—What Is It?

Khormun. (2018).*Rule of Sharjah inaugurates new Khorfakkan road*. Khorfakan City Municipality https://khormun.gov.ae/en/newsdetails/3295c76acbf4caaed33c36b1b5fc2cb1

Lombardi, D. R., Leach, J. M., Rogers, C.D.F., & the Urban Futures Team. (2012). *Designing resilient cities: A guide to good practice: (EP103)*. IHS BRE Press. https://www.brebookshop.com/details.jsp?id=326921

Net Zero Together Bootcamp—What Is It?—Birmingham Centre for Railway Research and Education (BCRRE) Blog. (2023, March 10). Retrieved August 2, 2024, from Bham.ac.uk website: https://blog.bham.ac.uk/bcrre/2023/03/10/net-zero-together-bootcamp-what-is-it/

PAS 2080:2023. (BSi, 2023). Carbon management in buildings and infrastructure.

Qiao, Y., Peng, F., Sabri, S., & Rajabifard, A. (2019). Low carbon effects of urban underground space. *Sustainable Cities and Society, 45*, 2210–6707.

Sadeghi, H., & Singh, R. M. (2023). Driven precast concrete geothermal energy piles: Current state of knowledge. *Building and Environment, 228*, 109790–109790. https://doi.org/10.1016/j.buildenv.2022.109790

UN. (2023). For a livable climate: Net-zero commitments must be backed by credible action. https://www.un.org/en/climatechange/net-zero-coalition#:~:text=What%20is%20net%20zero%3F,oceans%20and%20forests%20for%20instance.

UNFCC. (1997). The Paris Agreement. https://unfccc.int/process-and-meetings/the-paris-agreement?gad_source=1&gclid=Cj0KCQiAxOauBhCaARIsAEbUSQSuTKUzDtzkx8et_Tns1k1RZK_afPAchoEKjhjLDME7JvlMp2fLHFQaAuTtEALw_wcB

UNFCC. (2023). COP28. https://unfccc.int/cop28

Utility Tunnels (MUTs). Lecture Notes in Civil Engineering, pp. 149–162. https://doi.org/10.1007/978-981-19-1029-6_12

World Climate Action Summit Summary—COP28 UAE. (2023). Retrieved August 2, 2024, from Cop28.com website. https://www.cop28.com/en/world-climate-action-summit

Xia, B., Ziye, G., Peng, Z., & Xiaozhao, L. (2024). Quantitative analysis of low carbon effect of urban underground space in Xinjiekou district of Nanjing city, China. In: *Tunnelling and Underground Space Technology* (Vol. 143). Quantitative analysis of low carbon effect of urban underground space in Xinjiekou district of Nanjing city, China. ScienceDirect.

Xie, L., Li, Z., Li, J., Yang, G., Jiang, J., Liu, Z., & Tong, S. (2022). The impact of traditional raw earth dwellings' envelope retrofitting on energy saving: A case study from Zhushan Village, in West of Hunan China. *Atmosphere, 13*, 1537. https://doi.org/10.3390/atmos13101537

Chapter 2
Reverse Electrodialysis: An Energy Generation System from Desalination's Rejected Brine Disposal

Muhammad Wajid Saleem and Hassan Ali

Abstract Desalination plants' rejected brine which is hazard to the environment, marine life, and the agricultural land can be diluted through a novel approach of reverse electro-dialysis (RED). RED, which is a salinity gradient power generation system, will generate electricity due to salt concentration gradient between the rejected brine and sea water. The energy generated through RED will reduce the specific energy consumption of the desalinated water. In GCC, fossil fuel is used for electricity generation and water desalination, which produce greenhouse gases. Furthermore, due to the high operating cost in terms of fuel used, government subsidies the water desalination plants, which is an extra load on country's economy. Thus, the reduction in specific energy consumption for water desalination as proposed will affect the region's economy and environment positively.

Keywords Sustainable environment · Renewable energy generation · Rejected brine disposal · Salinity gradient power generation systems

Introduction

The wastewater of desalination plant known as brine or rejected water disposal is an economical and environmental issue. Generally, the brine rejected from the desalination plant has up to 98% TDS with concentration reaching 70,000 ppm in some cases (Ahmad & Williams, 2011; Ersever et al., 2007). Disposal of hyper concentrated brine improperly pollutes the groundwater and has an adverse impact on marine life. Agricultural land is also affected due to this concentrated water and cultivation stops at

M. W. Saleem (✉)
Faculty of Computing, Engineering and Media, School of Engineering and Sustainable Development, De Montfort University, Dubai, UAE
e-mail: Muhammad.saleem@dmu.ac.uk

H. Ali
School of Chemical Engineering, University of Birmingham, Dubai, UAE
e-mail: h.a.ali.1@bham.ac.uk

© The Author(s), under exclusive license to Springer Nature Switzerland AG 2026
J. Dsilva (ed.), *Redefining Sustainability*, https://doi.org/10.1007/978-3-031-88376-7_2

the specific farmlands. Various techniques are adopted to mitigate the brine disposal problem such as evaporation pond, deep well injection, disposal into surface water bodies, planation of trees that can tolerate the saline water, and conventional crystallizers (Shalaby et al., 2022). However, all these techniques have some constraints in terms of disposal cost, geographical location, quantity and quality of brine water, and feasibility to adopt in terms of government policies. This motivates us to find an alternative solution which will be potentially viable and economical. Brine dilution with energy generation from brine is an alternative option proposed in this abstract. Salinity gradient power (SGP), a renewable energy generation technique, is adopted for this purpose. In SGP system, energy generated due to the salt concentration difference between the salinity streams. An integration of SGP system with the desalination plant will utilize the rejected brine of desalination plant and generate electric energy. This integrated process reduces the overall energy consumption of the desalination system, since a percentage of the energy consumption for desalination process can be recovered through SGP technique. Reverse electro-dialysis (RED) is the promising SGP technology used to extract energy from salinity streams flowing alternatively in flow channels. RED cells are comprising of ion exchange membranes (IEM), electrodes, and spacer as shown in Fig. 2.1. Semi-permeable membranes of anion exchange membrane (AEM) and cations exchange membrane (CEM) are used alternatively to develop the channels. High concentration stream (HCS) and low concentration stream (LCS) flow through these channels to generate the electricity at the electrodes. It is reported that theoretically 0.8 kWh of energy can be generated with 1 m^3 of river water flow into the sea, which converts to 2 TW globally.

Different studies and processes are conducted for brine disposal (Shalaby et al., 2022). However, a common issue is observed in all techniques and that is the extra operating cost of brine disposal to the desalination plant. Therefore, an approach to develop a technique which instead of extra burden to the desalination plant, helps to reduce the overall operating cost of the plant should be considered. Integration of salinity gradient power with desalination process for energy generation and brine dilution is best suitable solution of problem. This is a novel approach of brine disposal, and no significant research work has been reported till now. Thus, evaluating the practical application of SGP process with desalination plant brine data is a promising effort.

The concept of SGP dates back to 1954 by Pattle known as blue energy (Pattle, 1954). But it did not perform significantly due to the constraint of poor performance of the membrane technology. However, in recent years, a lot of efforts are done to improve the power density of SGP systems by manufacturing Nano-pores and Nano-fluidic membranes (Tong et al., 2021). In 2014, the first demonstration unit of RED with 50 m^2 of membrane area for 125 cell pairs stack produce the power density of 1.3 W/m^2 with real brackish water samples (Tedesco et al., 2017). Similarly, power density of 4.3 W/m^2 was also generated through microbial RED cell unit. Furthermore, the power density of RED is directly proportional with the concentration gradient of salinity streams flowing through the channels of RED. The rejected brine of desalination plant has a high concentration gradient. Thus, the energy generation in terms of power density will also be high.

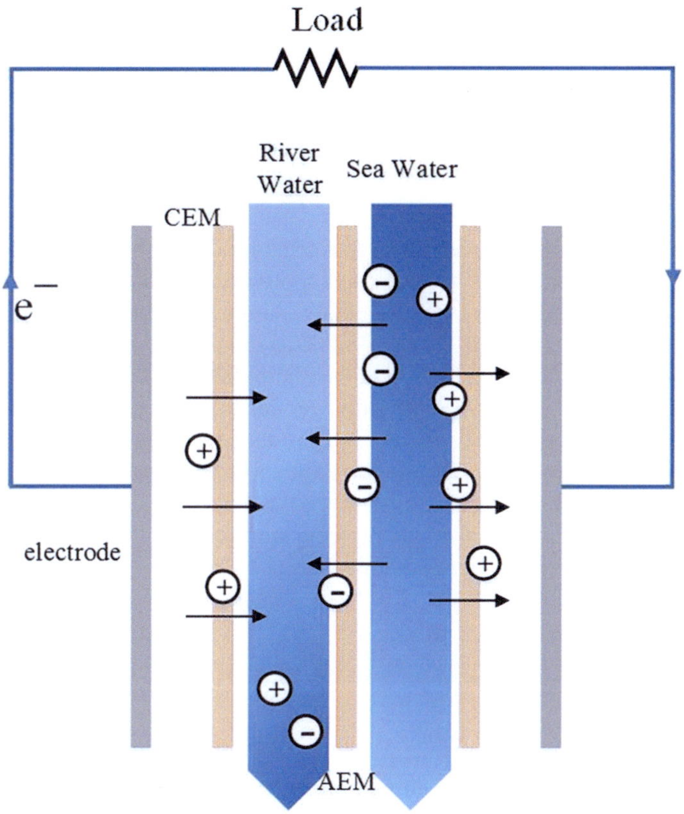

Fig. 2.1 Schematic diagram of reverse electro-dialysis (RED)

Methodology

In the Gulf, desalination projects take up the biggest chunk of the economy and the government spends a lot of money to make the freshwater accessible to everyone. The energy consumption of the desalination plant is a major portion of the operating cost of the plant. Therefore, experimentally evaluating an energy recovery mechanism as an alternative of previously available brine disposal methods is the most desirable approach. In this conference paper, we propose an idea to utilize the desalination plant rejected brine with the SGP simulation model to predict the overall specific energy consumption for freshwater generation.

There are basically two steps in this paper: data acquisition from desalination plant and mathematical model. The concentration of salt in rejected brine from the desalination plant in GCC is acquired. The data is then used in the mathematical model of RED to find out the energy generation while considering the real desalination plant brine water volume flowrate and concentration. The mathematical model of RED is prepared by using the MATLAB software and already published in our previous papers (Saleem et al., 2017).

Result and Discussion

It was observed from the collected data that brine produced through the desalination plants of UAE has concentration increased approximately 8000–10,000 ppm when compared with the feed water for these plants (Amma & Ashraf, 2020). The salinity concentration in gulf is around 35,000–48,000 as reported by UAE Ministry of Environment and Water. According to the RED mathematical model and the concentration gradient given above, about 0.9 W/m^2 of power density can be produced through this salinity gradient process. Since the energy produced is very little, this technology will be beneficial in terms of mitigate the issue of brine disposal without increasing the desalination plant operational cost. Because the energy produced through this RED process will reduce the plant energy consumption cost that can compensate for the RED membrane replacement and installation cost.

Conclusion

In conclusion, the integration of Reverse Electro-Dialysis (RED) as a salinity gradient power generation system with desalination plants offers a promising solution to the environmental and economic challenges associated with rejected brine disposal. By harnessing the energy generated from the salt concentration gradient between rejected brine and seawater, this approach not only reduces the specific energy consumption of desalination but also provides a sustainable means of disposing of brine without increasing operational costs. The potential of this technology, coupled with ongoing advancements in membrane technology and power density improvements, makes it a compelling option for the Gulf region's desalination plants, contributing to both economic and environmental sustainability.

References

Amma, L. V., & Ashraf, F. (2020, February). Brine management in reverse osmosis desalination: A UAE perspective. In *2020 Advances in Science and Engineering Technology International Conferences (ASET)* (pp. 1–6). IEEE.

Ahmad, M., & Williams, P. (2011). Assessment of desalination technologies for high saline brine applications—Discussion paper.*Desalination & Water Treatment, 30.*

Bindels, M., Carvalho, J., Gonzalez, C. B., Brand, N., & Nelemans, B. (2020). Techno-economic assessment of seawater reverse osmosis (SWRO) brine treatment with air gap membrane distillation (AGMD). *Desalination, 489*, 114532.

Ersever, I., Ravindran, V., & Pirbazari, M. (2007). Biological denitrification of reverse osmosis brine concentrates: I. batch reactor and chemostat studies. *Journal of Environmental Engineering and Science, 6*(5), 503–518.

Pattle, R. E. (1954). Production of electric power by mixing fresh and salt water in the hydroelectric pile. *Nature, 174*(4431), 660.

Shalaby, S. M., Sharshir, S. W., Kabeel, A. E., Kandeal, A. W., Abosheiasha, H. F., Abdelgaied, M., … Yang, N. (2022). Reverse osmosis desalination systems powered by solar energy: Preheating techniques and brine disposal challenges–A detailed review. *Energy Conversion and Management, 251*, 114971.

Saleem, M. W., Jande, Y. A. C., & Kim, W. S. (2017). Performance optimization of integrated electrochemical capacitive deionization and reverse electrodialysis model through a series pass desorption process. *Journal of Electroanalytical Chemistry, 795*, 41–50.

Tang, W., & Ng, H. Y. (2008). Concentration of brine by forward osmosis: Performance and influence of membrane structure. *Desalination, 224*(1–3), 143–153.

Tedesco, M., Cipollina, A., Tamburini, A., & Micale, G. (2017). Towards 1 kW power production in a reverse electrodialysis pilot plant with saline waters and concentrated brines. *Journal of Membrane Science, 522*, 226–236.

Tong, X., Liu, S., Crittenden, J., & Chen, Y. (2021). Nanofluidic membranes to address the challenges of salinity gradient power harvesting. *ACS Nano, 15*(4), 5838–5860.

Chapter 3
Chronological Development of NZEB: A Comprehensive Review

Ehtesham Matin and Mohd Khalid Hassan

Abstract The ascent of Net Zero Energy Buildings (NZEBs) as a key climate change solution is intricately linked to a convergence of global factors and significant world events. In the twenty-first century, where energy is a linchpin for the functioning of economies and nations, the concept of self-sustainability has evolved into a paramount objective. This evolution has witnessed diverse concepts and advancements in building design and energy efficiency, prompting continual refinement of NZEB definitions and conceptual frameworks. Discerning the pivotal role of energy on a global scale, there is a compelling need for a standardized international framework for NZEBs, given the ongoing evolution of definitions over decades. Noteworthy is the importance of definitions tailored to local conditions, striking a crucial balance between an internationally standardized framework and locally adaptable definitions. Recognizing the compelling need for standardization, this paper aims to chronicle the chronological development of NZEB definitions and frameworks. The primary objectives include comprehensive understanding and revealing the diverse parameters inherent to NZEB by reviewing various definitions and studies. The goal is to contribute to the development of a standardized framework that effectively guides NZEB implementation. This involves striking a balance between internationally standardized frameworks and locally adaptable definitions, acknowledging the relevance of tailoring definitions to local conditions and emphasizing the importance of a comprehensive understanding of NZEB parameters.

Keywords Net Zero Energy Buildings (NZEB) · Standardized framework · Energy efficiency · Building performance · Self-sustainability

---

E. Matin (✉)
Department of Architecture and Planning, IIT Roorkee, Roorkee, India
e-mail: ehtesham_m@ar.iitr.ac.in

M. K. Hassan
Department of Architecture, ZHCET, AMU, Aligarh, India
e-mail: arch5.khalidhasan@gmail.com

© The Author(s), under exclusive license to Springer Nature Switzerland AG 2026
J. Dsilva (ed.), *Redefining Sustainability*, https://doi.org/10.1007/978-3-031-88376-7_3

Introduction

Net Zero Energy Buildings (NZEBs) have emerged as a pivotal solution to mitigate the environmental impact of buildings while addressing the escalating energy demand (Singh et al., 2021), considering that buildings are responsible for a substantial 40% of global energy consumption (Delmastro & Chen, 2022). These structures are engineered for high efficiency, featuring low energy demand and the capability to fulfil all operational energy needs, all while ensuring optimal thermal and visual comfort (Saini et al., 2022) Apart from their economic and environmental advantages, NZEBs contribute significantly to energy security and resilience during power outages and natural disasters (Shetty et al., 2023).

Over the past century, the definitions and concepts related to energy-efficient buildings have undergone constant evolution, reflecting the changing dynamics of the global order. This evolution has witnessed a spectrum of multifaceted ideas, ranging from solar houses relying on passive techniques to the high-end technological development of various active systems, all aiming to declare the energy efficiency of built systems through the integration of renewable energy sources. These designs and engineering solutions are intricately crafted for specific contexts and needs, posing a challenge when it comes to evaluating their performance on a standardized scale.

Acknowledging the ever-changing landscape of energy-efficient building concepts and the diverse approaches adopted over time, there is a need to navigate through this complexity. The urgency to address the increasing significance of energy within contemporary architectural practices underscores the necessity for establishing a globally standardized framework for NZEBs. However, such a framework must maintain flexibility to accommodate the intrinsic efficacy of contextually specific definitions (Deng et al., 2014; Mlecnik et al., 2011). This adaptability ensures that the framework is tailored to encompass the diverse climatic, socio-economic, and infrastructural nuances inherent in specific regions. By striking this delicate balance, the envisioned framework can effectively guide the evaluation and implementation of NZEB principles, fostering a harmonious integration of energy-efficient technologies across the global architectural landscape. Therefore, achieving a harmonious equilibrium between the universality of a standardized framework and the granularity of localized definitions becomes a fundamental imperative. This balance is crucial for effectively addressing the multifaceted needs of diverse geographical landscapes in the pursuit of sustainable and energy-efficient building practices. Motivated by the ongoing quest for net zero energy goals in the built environment, this research embarks on a meticulous investigation into the diachronic evolution of Net Zero Energy Building (NZEB) conceptualizations and operational frameworks.

The overarching objective of this study is to elucidate the fundamental parameters intrinsic to authenticating NZEB design and performance. Through this rigorous inquiry, the aim is to identify a globally applicable, standardized framework that demonstrably steers the efficacious implementation of NZEB principles. The framework posits clarity in NZEB definitions as a cornerstone element, aiming to expedite the streamlining of energy-efficient regulations and foster vibrant knowledge exchange within the construction sector. In doing so, this study seeks to contribute to a more sustainable and resilient built environment, aligned with the global imperative for responsible energy consumption and environmental stewardship.

Methodology

The research philosophy shall be pragmatic and based on secondary sources. The data shall be collected by a literature review of books, research papers, journals, standards and codes of practice. To grasp the concept, definition, and framework of Net Zero Energy Buildings (NZEB), an extensive search was conducted using Google Scholar. Keywords such as "Net Zero Energy Building," "definitions," "NZEB and thermal comfort," and "elements and framework of NZEB" were employed, resulting in the selection of 33 highly relevant studies spanning the period from 1970 to 2022. This would be helpful to chronicle the chronological development of NZEB definitions and frameworks over time and subsequently develop an international standardized framework.

Historic Evolution of Net Zero Energy Buildings

The fusion of Architecture and Energy has forged a harmonious synergy, giving rise to the paradigm of Net Zero Energy Buildings (NZEBs). This evolution is marked by a cascade of ground-breaking concepts, definitions, and advancements, heralding a new era of sustainable innovation. Beginning with the inception of solar energy utilization through the construction of the "*MIT Solar House I*" in 1939 (Levy, 2017), a trail of innovations and strategic shifts has led to the contemporary pursuit of buildings that harmoniously balance energy consumption and production. This also marked the commencement of an era where the notion of "*Passive Solar building*" emerged in response to the energy scarcities witnessed during World War 2 in the United States (Lewis, 2007). Architects like George Fred Keck and Louis Kahn pioneered the "*Solar House Principle*," employing design strategies such as southern-facing living areas, glass facades, roof overhangs, and insulation to optimize solar energy utilization in buildings (Barber, 2023). Concurrently, Bliss (1955) made contributions to solar air collectors and rock mass storage for solar-heated housing (Bliss, 1955). The active solar systems initially held prominence, from the 1950s to 1970s, for utilizing solar energy. However, a transition towards passive solar heating gained momentum due to its economic efficiency, minimal maintenance requirements, and heightened dependability (Hayes & Andrejko, 1983).

The oil crisis of the 1970s acted as a catalyst, driving the resurgence of *Solar Architecture*. This resurgence saw the development of innovative designs such as the *Zero Energy Home* in Denmark, featuring enhanced insulated envelopes, heat recovery systems, and integrated solar collectors on the southern façade (Esbensen &

Korsgaard, 1977). Additionally, noteworthy examples like the *Balcomb house* characterized by its high thermal mass, Trombe wall, and an active air circulation system for thermal comfort (Balcomb et al., 1980). Other notable projects that emerged during this era include the Saskatchewan *Conservative House* in Canada (Besant et al., 1979) and the *Natural Energy Autonomous House* in Japan (Saitoh, 1984).

A milestone was reached in 1990 with the *Passivhaus* Standard when a row of townhouses was constructed in Darmstadt Kranichstein, Germany, serving as proof of the concept. These structures prioritized superinsulation and airtightness in their design (Katzenbach et al., 2014). Similarly, the "*Self-Sufficient Solar House*" in Freiburg gained popularity for its half-cylindrical design, featuring an array of renewable energy systems such as air-to-ground heat exchange, ventilation heat recovery, solar hot water, transparent insulation, fuel cells, and a hydrogen tank (Stahl et al., 1994).

The declaration of the Intergovernmental Panel on Climate Change (IPCC) First Assessment Report in 1990 played a crucial role in identifying human activities, particularly the burning of fossil fuels and deforestation, as the main cause of Global Warming. This had a profound impact on the scientific community and led to the formal definitions of Net Zero Energy Buildings (NZEB). Gilijamse (1995) characterized a "*Zero Energy House*" as one ***devoid of fossil fuel consumption***, with "***annual electricity consumption equating to annual electricity production***," (Gilijamse, 1995). Iqbal (2003) subsequently augmented this definition by emphasizing the amalgamation of ***renewable energy technologies*** and cutting-edge ***energy-efficient construction techniques*** (Iqbal, 2003). This period also witnessed a shift in focus from residential housing to diverse building typologies.

Torcellini et al. (2006) proposed a comprehensive definition for a Zero Energy Building (ZEB), advocating for an initial emphasis on energy efficiency followed by the integration of renewable energy sources. This approach laid the groundwork for establishing the concept of Net Zero Energy Buildings (NZEBs) and centring around four crucial metrics: "*Net Zero **Site**Energy,*" "*Net Zero **Source**Energy,*" "*Net Zero Energy **Cost**,*" and "*Net Zero Energy **Emission**.*" Within this framework, they also categorized renewable energy supply into five distinct options, which included *enhancements in energy efficiency*, utilization of ***on-site****building footprint* and at-the-site renewable energy resources, sourcing energy from **off-site** locations, and the *purchasing* of renewable energy from external sources (Torcellini et al., 2006). Marszal et al. (2011) further delineated these categories using Roman numerals for various boundaries of RE supply options, including ***Production on building footprint (I)***, On-site production from on***-site renewables (II)***, On-site production from ***off-site renewables (III)***, ***Off-site production (IV)*** and ***Off-site purchase (V)*** (Marszal et al., 2011) (Fig. 3.1).

Expanding upon this foundation, Pless and Torcellini (2010) introduced a comprehensive classification system rooted in strategies for renewable energy supply. This classification system encompassed various NZEB types, each with distinct characteristics. These included NZEBs where energy demand is met within the building footprint **(NZEB: A)**, those fulfilled on the building site **(NZEB: B)**, those relying

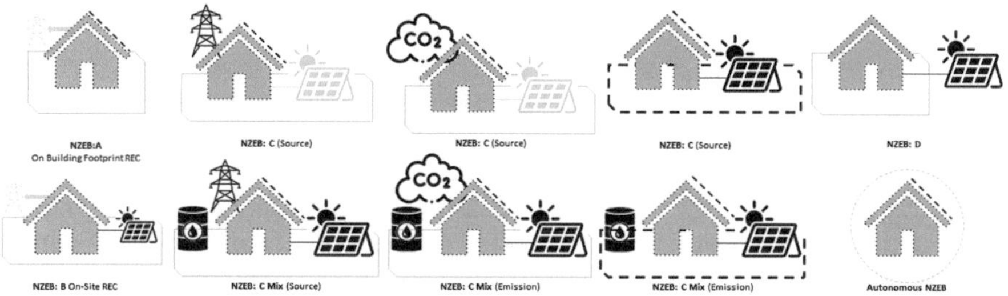

Fig. 3.1 Renewable supply options (*Source* Marszal et. al., 2011)

on off-site renewable sources **(NZEB: C)**, and those depending on energy imports from externally sourced renewable resources **(NZEB: D)** (Pless & Torcellini, 2010).

Building upon these classifications, Kaewunruen et al. (2019) further refined the NZEB: C category. They introduced a more detailed evaluation framework that considered Site, Source, and Emission aspects. In doing so, they introduced two additional categories, NZEB Mix (Site) and NZEB Mix (Source), which placed specific emphasis on particular dimensions of renewable energy integration within the NZEB concept. In a significant addition, they introduced the concept of Autonomous NZEB, expanding the spectrum of NZEB definitions to a total of ten distinct categories (Kaewunruen et al., 2019).

In 2007, Kilkis brought attention to a crucial distinction between "*Net Zero Energy Buildings*" and "*Net Zero Carbon Buildings*". These terms, while related, do not convey the same concept. The former denotes buildings capable of generating as much energy as they consume, while the latter pertains to the offsetting of carbon emissions. Kilkis emphasized limitations in the conventional definition of Net Zero Energy Buildings (NZEBs) as it primarily focuses on quantifying energy savings and generation, neglecting the crucial aspect of energy quality, referred to as "exergy". To address this conceptual gap, Kilkis introduced a novel metric, the "***Net Zero Exergy Building***" (ZEXB). A ZEXB is defined by its ability to achieve a balance in exergy exchange over a year at the building boundary within an energy system. This metric encompasses different forms of energy transfer occurring over a specific timeframe. The ZEXB concept aims to provide a more comprehensive and nuanced evaluation of a building's energy performance by considering not only the quantity of energy but also its quality and efficiency in contributing to the broader energy landscape (Kilkis, 2007). EPDB introduced the term "***nearly Zero Energy Building (nZEB)***," describing a building with high energy performance predominantly fulfilled by renewable sources. Various approaches to metrics have led to diverse classifications for Nearly Zero Energy Buildings (NZEBs) (EPDB, 2010).

Considering the embodied energy of building elements, Hernandez and Kenny (2010) introduce the concept of ***"Life Cycle Zero Energy Buildings"*** (LC-ZEB), coupled with an annual energy consumption (Hernandez & Kenny, 2010). Cole and Fedoruk (2014) introduced the concept of "***Net Positive Energy Buildings***" (NPEBs) and explained how they aim to surpass the traditional approach of "Net Zero Energy

Buildings" (NZEBs) by not only producing enough energy for their use but also generating surplus energy that can be shared with the grid or utilized elsewhere. It underscores the comprehensive perspective of NPEBs, encompassing elements such as the complete lifespan of a building, the quality of generated energy, and the incorporation of social processes within energy systems (Cole & Fedoruk, 2014).

Taherahmadi et al. (2021) presented an intricate classification system based on yearly or lifecycle periods, encompassing Energy, Carbon, Exergy, and Cost dimensions, and accounting for on and off grid connections. This comprehensive framework introduces a spectrum of 28 definitions, encompassing three distinct variations for each on-grid connection type. These definitions span positive, negative, and balanced or net dimensions, offering a holistic perspective on the diverse ways NZEBs can be defined and understood (Taherahmadi et al., 2021). In the context of on-grid connectivity, NZEBs strive to fulfil the entire energy demands of the building using renewable energy sources situated on the premises, with the flexibility to draw power from the nearby electrical grid as necessary. If on-site renewables generate an energy surplus beyond the building's requirements, the excess can be contributed to the local grid. (Jaysawal et al., 2022). Buildings play a crucial role in shielding us from extreme weather conditions. To do so effectively, it is imperative to strike a balance between maintaining optimal thermal comfort and the efficient use of energy. At the same time, the preservation of indoor air quality remains a top priority, without compromising the specific functions a building serves. Inevitably, these operations require a significant amount of energy, and this demand can be met by integrating renewable energy systems. In essence, a NZEB is designed to provide "all essential operational energy along with thermal comfort and indoor environment quality within the optimized energy demand which may be fulfilled through a renewable energy source" (Saini et al., 2022). Reflecting on decades of research, Omrany et al. (2022) highlight the evolution of NZEB design. Initially, the focus was primarily on reducing the environmental impact of buildings. However, this approach has since transformed into a more concentrated effort to enhance energy performance. This progression encompasses various aspects, including energy conservation, energy production, energy storage, and innovative methodological approaches to building design (Omrany et al., 2022).

The IPCC's Fourth Assessment Report (AR4) in 2007 triggered global action on human-induced warming, culminating in the Paris Agreement of 2015, where Net Zero Energy Buildings (NZEBs) gained prominence in national strategies. Countries like Germany and France integrated NZEB targets into building codes, supported by initiatives like the International NZEB Alliance. The IPCC's Fifth Assessment Report (AR5) in 2013–2014 underscored the urgency of emissions cuts, driving the NZEB revolution. Technological innovations in renewable energy and materials enhanced NZEB feasibility, while AR5 emphasized tailored design for diverse climates. These developments propelled NZEBs as key solutions in combating climate change (IEA, 2021).

Framework for NZEB Definition

A standardized framework for defining "net zero energy buildings" (NZEBs) is pivotal in establishing precise and consistent criteria for these energy-efficient structures. This framework encompasses five key components that shape the definition and evaluation of NZEBs. Firstly, it delineates a ***system boundary***, considering both physical and economic factors, to determine the aspects of energy use and production to be included. Secondly, a ***weighting system*** is introduced to convert energy measurements into comprehensive metrics, facilitating a holistic assessment of the entire energy lifecycle. Thirdly, achieving a ***net zero energy balance*** necessitates evaluating energy use and generation over different timeframes, emphasizing energy efficiency, indoor comfort, and renewable energy sources. Fourthly, ***temporal energy interplay***, synchronization of energy consumption, and production within a building over time to optimize efficiency and grid interaction. Finally, ***energy forecasting and robust monitoring***, utilizing measured or calculated ratings, validate NZEB status and ensure sustained performance while maintaining indoor comfort and compliance.

Such a framework simplifies NZEB classification, enhances regulatory understanding, and promotes knowledge exchange for constructing highly efficient buildings (Deng et al., 2014; Kurnitski et al., 2011). It is important to note that definitions reflecting local conditions, such as climate, customs, and economic factors, are more appropriate (Deng et al., 2014; Mlecnik et al., 2011), as there are various building types with different purposes and requirements, and a single definition may not capture this diversity (Marszal & Heiselberg, 2009). Various NZEB definitions depend on regional factors, government energy efficiency goals, user needs, and the aspiration to achieve zero energy. (Pless & Torcellini, 2010; Sartori et al., 2012). Moreover, these definitions can evolve to adapt to changing energy needs and consumption patterns (Saini et al., 2022). Overall, clear and standardized NZEB definitions are crucial for simplifying the understanding of energy-efficient regulations and fostering the exchange of knowledge on constructing highly efficient buildings (Kurnitski et al., 2011).

Building System Boundary

Defining the system boundary for Net Zero Energy Buildings (NZEBs) is crucial. This boundary serves as a point of reference to assess the inflow and outflow of energy within the system, taking into account both physical and balance-related considerations (Kurnitski et al., 2011; Sartori et al., 2012). The building system boundary can extend beyond the physical building footprint to include virtual economic aspects (Deng et al., 2014). It is instrumental in determining which aspects of energy consumption and production are to be included in the calculations, encompassing various factors such as heating, cooling, lighting, appliances, hot water, and servers(Attia, 2018; Kurnitski et al., 2011).

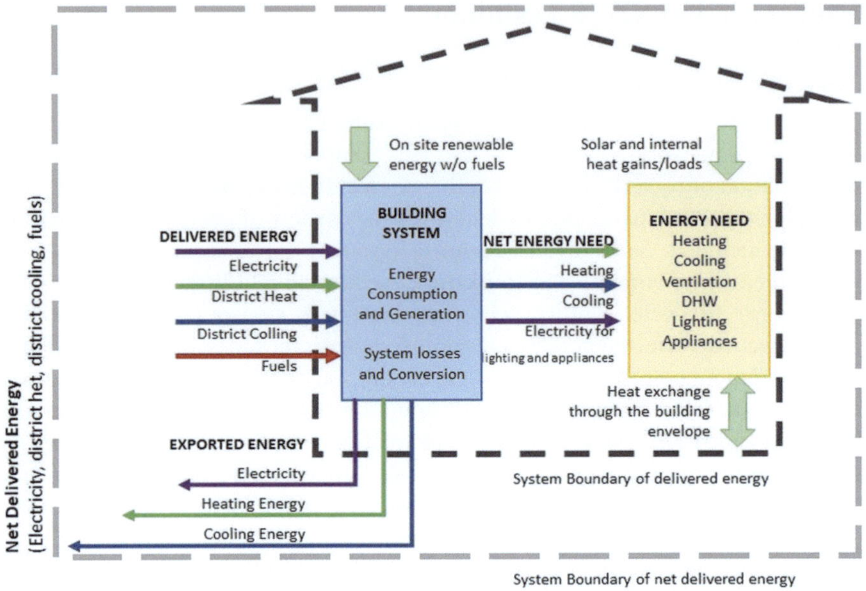

Fig. 3.2 System boundary for net delivered energy (*Source* Kurnitski et al., 2011)

The boundary delineates the energy need and subsequent building of technical systems for energy use and production (Kurnitski et al., 2011) and identifies the renewable energy generation that contributes to delivered and exported energy through the power grid (Sartori et al., 2012). The wider boundary allows for various renewable energy supply options, including the purchase of "green electricity" from large-scale renewable power plants, and helps classify the type of renewable energy system as on-site or off-site (Deng et al., 2014).

In essence, this boundary covers the entire spectrum of energy demand within a building, encompassing operational energy needs like heating, cooling, and electricity, as well as considerations of embodied energy in construction materials and energy associated with occupancy (Marszal & Heiselberg, 2009). This delineation of boundaries can be applied to individual structures or groups of buildings, taking into consideration factors like district heating, efficiency measures, and the types of fuels used (Sartori et al., 2010) (Fig. 3.2).

Physical Boundary

The physical boundary in Net Zero Energy Buildings (NZEBs) applies to single or grouped buildings, differentiating on-site from off-site energy generation and addressing ownership of renewable installations (Sartori et al., 2012). Establishing precise boundaries definition enables a comprehensive evaluation of energy sources and consumption within the building context (Marszal & Heiselberg, 2009; Voss et al., 2011).

Balance Boundary

The balance boundary in Net Zero Energy Buildings (NZEBs) determines which energy uses are counted in the NZEB balance. Typically, it covers *operational energy* like heating, cooling, lighting, and more as per standards. Net ZEB assessments can also include nonstandard energy uses like rainwater treatment or electric vehicle charging. Moreover, they consider embodied energy and emissions from materials and installations, broadening Net ZEBs' environmental scope. To account for *embodied energy* and emissions, annualization is needed, based on assumptions about the building's lifespan and component durability. While energy used during construction and demolition could be considered, it is usually minor and often disregarded (Jaysawal et al., 2022; Saini et al., 2022; Sartori et al., 2012).

Boundary Conditions

Functionality and Space Effectiveness

The building's functionality specifies the occupancy level. The density and occupancy significantly influence the building's energy needs. Additionally, specifying the building type, whether it is a detached house, apartment, office, school, or falls into another category, is fundamental for making meaningful comparisons with similar structures. These metrics not only facilitate comparisons with similar buildings but also help evaluate whether monitored data aligns with design expectations (Marszal & Heiselberg, 2009; Sartori et al., 2010, 2012).

Climate and Comfort

Identifying the reference climate and the comfort standards employed in load calculations is crucial for comparisons with similar buildings and for cross-referencing monitored and design data, as variations in climate conditions and user behaviour can significantly impact energy demand (Deng et al., 2014; Mlecnik et al., 2011; Sartori et al., 2010).

Weighting System

The Weighting System in Net Zero Energy Building (NZEB) Assessment ensures the comprehensive evaluation of energy metrics. This system facilitates the conversion of energy measurements into diverse metrics, considering factors like energy extraction, generation, and delivery, as well as political considerations. It provides a consistent assessment of the entire energy lifecycle, accounting for natural energy sources, conversion processes, and distribution grids (Sartori et al., 2012). Moreover, it accommodates the fuel-switching effect and necessitates regular updates to align with grid developments (Sartori et al., 2010). The framework explores various ***metrics***, including primary energy, CO_2 equivalent emissions, and energy cost, to define the equilibrium of energy and emissions, considering regional influences and the chosen balancing period. The ***balancing period***, whether annual or extended over a building's lifecycle, impacts how NZEB performance is evaluated. Additionally, it discusses ***symmetric*** and asymmetric weighting approaches and emphasizes the importance of ***time-dependent accounting*** for optimizing building-grid interaction within evolving energy infrastructures.

Balancing Metric

The "Balancing Metric" serves as a measurement criterion for assessing the equilibrium or balance of energy and emissions (Attia, 2018). There is a wide range of metrics that has been used to determine this "zero" balance, including final energy, primary energy, CO_2 equivalent emissions, exergy, energy cost, or metrics guided by national energy policies. These definitions can vary depending on project objectives, investor intentions, climate considerations, and economic factors. Commonly employed metrics include site energy, source energy, emissions, and cost (Torcellini et al., 2006). Notably, primary energy emerges as the most prevalent metric for assessing the net ZEB balance (Marszal & Heiselberg, 2009; Marszal et al., 2011; Sartori et al., 2010) Primary energy can encompass renewable, non-renewable, or hybrid sources, adding to its versatility (Voss et al., 2011). The selection of metrics and conversion factors can be influenced by political considerations, leading to preferences for specific energy sources and technologies. This, in turn, can impact electricity generation requisites and steer the adoption of particular solutions. For instance, solar energy might gain favour over biomass in regions where biomass availability is limited due to non-energy uses such as food production (Sartori et al., 2012).

Balancing Period

For energy calculation, annual balance is widely considered due to its adaptability to varying operational scenarios and seasonal fluctuations (Marszal et al., 2011) However, this method relies on typical meteorological data, potentially overlooking extreme years and resulting in variable energy budgets (Voss et al., 2011). Conversely, shorter periods, like monthly or seasonal balances, present design and assessment challenges (Marszal et al., 2011). Researchers like Hernandez and Kenny (2010) propose a comprehensive perspective encompassing a building's entire lifecycle. This extended timeframe considers not only operational energy but also embodied energy from construction, demolition, and installations, offering a holistic view spanning decade (Hernandez & Kenny, 2010). Notably, embodied energy, often underestimated, can account for 20–30% of total primary energy expenditure for energy-efficient buildings with long lifespans (Voss et al., 2011). To assess Net Zero Energy Building performance comprehensively, embodied energy can be annualized and integrated with operational energy. Ultimately, the choice of a yearly balance remains fundamental for evaluating net energy performance. However, it allows flexibility to consider alternative timeframes, including multi-year or seasonal balances, particularly during extensive renovation periods spanning decades. This adaptability recognizes the intricate interplay between energy efficiency measures and supply systems across a building's dynamic lifecycle (Sartori et al., 2012).

Symmetry

Symmetric weighting and asymmetric weighting are two approaches to assessing the value of exported and received energy in different scenarios. Symmetric weighting treats both forms of energy equally, assuming that on-site generation doesn't create problems elsewhere. For instance, with on-site cogeneration, although it increases delivered energy, it still allows full credit for exported electricity. On the other hand, asymmetric weighting comes into play when on-site generation has unaccounted negative effects. For instance, with PV generation, if embodied energy is considered in the balance, we can fully credit exported electricity; otherwise, some energy cost may not be credited. This approach enables differentiation between generation technologies like PV and cogeneration. The choice between symmetric and asymmetric weighting depends on various factors, including energy costs, transportation losses, and grid expenses. Symmetric weighting is suitable for scenarios where self-consumption of on-site generation is preferred, while asymmetric weighting can be used to promote technology adoption, such as early use of PV solar panels, by giving more weight to exported energy, which is compensated at a higher rate than delivered electricity (Sartori et al., 2012).

Time-Dependent Accounting

Time-dependent accounting in energy is complex but crucial. While static averages are common for energy and emissions, they may miss grid changes. Factors like electricity are often assessed regionally, while local sources are considered. Regular updates reflecting grid shifts are essential, and planning for future changes is vital. While hourly assessments are an option, they add complexity. It's better to use static or semi-static values for Net Zero Energy Building (NZEB) calculations and separately address real-time energy balance for practical building-grid optimization (Sartori et al., 2012).

Net Zero Balance

The foundational principle of Net Zero Energy Buildings (NZEBs) centres on achieving equilibrium between energy consumption and supply, allowing surplus credits to be exported to the Energy Infrastructure Grid within a defined timeframe, typically a year (Sartori et al., 2010; Voss & Musall, 2013). This equilibrium is systematically attained through a strategic process involving the following steps.

(i) ***Reduction of Energy Demand***: The initial focus is on reducing overall energy demand through the implementation of Energy-Efficient Measures. These measures not only elevate occupant comfort but also contribute significantly to emission reduction.

(ii) ***Integration of Renewable Energy Systems (RES)***: Following the reduction in energy demand, Renewable Energy Systems (RES) are systematically integrated to offset these demands. The emphasis is on leveraging on-site renewable sources, a critical step in achieving Net Zero Energy (NZEB) status for buildings. Additionally, surplus energy generated can be efficiently redirected back into the grid (Voss et al., 2010)

(iii) ***Intelligent Energy Grid Infrastructure***: The incorporation of Intelligent Energy Grid Infrastructure, encompassing smart grids and green power grids, plays a pivotal role in achieving net zero consumption while concurrently enhancing a building's resilience to grid failures.

This systematic approach not only underscores the technical intricacies of NZEBs but also emphasizes the interplay of various elements to attain a state of energy equilibrium and sustainability in building operations.

Achieving this balance necessitates rigorous evaluations of energy balance calculations, encompassing relevant indicators, clearly defined boundaries of the energy balance, and specification of the designated balance period (Feng et al., 2019; Voss & Musall, 2013; Wu & Skye, 2021). One vital aspect of attaining this balance is the forecast of building performance, which is a challenging task due to factors such as occupant behaviour and the complexities of active energy systems (Crawley et al.,

2009). Energy modelling is a valuable tool to estimate overall energy consumption, determine the necessary capacity for on-site renewable energy sources, and pinpoint opportunities for high-impact enhancements in energy efficiency. Achieving a Net Zero Energy (NZEB) status for buildings can be possible by focusing on substantial on-site renewable energy generation, potentially reducing the need for extensive energy efficiency measures (Feng et al., 2019).

Energy Efficiency and Indoor Environment Quality

Energy efficiency stands as the cornerstone of Net Zero Energy Buildings (NZEBs), encompassing a multifaceted strategy to optimize energy consumption across various functions like space heating, cooling, lighting, and appliances (Attia, 2018; Moghaddasi et al., 2021). It is not just about reducing energy consumption; it is about achieving NZEBs' primary goal—minimizing energy usage. This approach isn't merely cost-effective; it is a profound step towards reducing carbon emissions and bolstering building resilience, aligning perfectly with sustainability objectives (Jaysawal et al., 2022). The beauty of energy efficiency lies in its scalability, offering improved indoor comfort, elevated property value, and regulatory compliance (Sartori et al., 2010). At its core, setting a fixed maximum energy use serves as a crucial starting point (Marszal & Heiselberg, 2009).

Reducing energy demand through energy-efficient measures forms the bedrock of NZEBs, substantially curbing energy needs. This reduction can be masterfully achieved through two approaches: Passive and Active Design Measures. ***Passive Design Measures*** encompass a spectrum of elements, from building envelope enhancements to thermal insulation, window glazing, shading, and airtightness. Meanwhile, ***Active Design Measures*** optimize energy efficiency within building energy systems like HVAC systems, efficient lighting, plumbing fixtures, and appliances. These measures not only reduce consumption and emissions but also elevate occupant thermal comfort, indoor environment quality, and functionality (Healy et al., 2015; Saini et al., 2022; Singh et al., 2021; Wu & Skye, 2021).

The logical progression in the quest for NZEBs involves prioritizing energy efficiency improvements before venturing into renewable energy initiatives. This strategy typically yields a superior return on investment (Jaysawal et al., 2022). Identifying national or commercial standards for minimum energy efficiency requirements is pivotal, and economic life cycle assessments play a crucial role in assessing cost-optimal energy performance levels. These assessments encompass various measures like building insulation, HVAC efficiency, and lighting to reduce energy demand. Mandatory energy efficiency requirements can take prescriptive or performance-based forms in NZEB definitions, often emphasizing cost optimization or demand reduction (Sartori et al., 2012). ***The Energy Use Intensity (EUI)*** or ***Energy Performance Index (EPI)*** emerges as a critical metric for comparing energy consumption in NZEBs, facilitating energy efficiency benchmarking. It is defined as the total energy consumption by a building in a year per square metre of area (Saini et al., 2022).

Moreover, a well-defined criterion for indoor environmental quality, particularly thermal comfort, significantly influences a building's energy performance. The choice of a thermal comfort model profoundly impacts energy performance assessments, taking into account variations in reference conditions for indoor environments (Attia & Carlucci, 2015). Prioritizing energy efficiency remains sustainable throughout a building's lifecycle, offering practicality and cost effectiveness in saving energy rather than generating additional energy (Marszal & Heiselberg, 2009).

Energy Supply Options

The core principle of a zero energy building (ZEB) lies in its independence from fossil fuels, relying solely on renewable energy sources (Marszal & Heiselberg, 2009). However, the journey to achieve Net Zero Energy Buildings (NZEBs) is not solely about electricity generation and export; it represents a holistic approach that revolves around optimizing electricity usage and significantly reducing overall energy consumption, while simultaneously offsetting it through a Renewable Energy System (RES) (Voss et al., 2011).

The RES encompasses all renewable energy sources that a building can harness to meet its annual energy requirements. It's an intricate integration of environmentally friendly and sustainable energy sources that underpin the concept of NZEBs. This synergy demands the seamless coalescence of Passive Design Measures and the Renewable Energy system (Healy et al., 2015; Saini et al., 2022; Singh et al., 2021).

While setting a minimum percentage of renewable energy to fulfil a building's energy needs is one approach, a more comprehensive strategy involves categorizing energy supply options. This strategic outlook prioritizes sources with minimal emissions, limited transportation losses, long term availability, and scalability. On-site solutions like solar panels and combined heat and power (CHP) systems take precedence over off-site alternatives (Sartori et al., 2012; Torcellini et al., 2006).

Energy Grid Infrastructure

To truly succeed in the quest for NZEBs, it is vital to see the electricity grid as more than just a passive receiver of energy; it serves as a dynamic virtual buffer. This perspective accentuates the significance of intelligent grid-connected load management (Gilijamse, 1995; Madathil et al., 2021). Building-grid interactions constitute the very heart of achieving a net energy balance (Torcellini et al., 2006), ensuring that a building's power imports and exports to the grid are finely balanced (Singh et al., 2021).

In this transformative journey, policymakers hold a key role. Their task is to discern between "supply-side" renewable generation, which can be sold and exported, and "demand-side" generation, focused on reducing energy consumption within buildings (Marszal & Heiselberg, 2009; Sartori et al., 2012). An effective strategy involves utilizing weighting factors to promote sustainability while considering local factors and a diverse range of supply options. This well rounded approach ensures a robust energy supply system that aligns seamlessly with NZEB objectives (Sartori et al., 2010). Importantly, achieving NZEB performance not only contributes to sustainability but also bolsters a building's resilience to grid failures (Shirinbakhsh & Harvey, 2021).

For an even greater impact, we can look towards green power grids, powered exclusively by renewables, as a clear pathway to achieving climate neutrality. These grids effectively decouple energy consumption from emissions, bringing us closer to our sustainability goals (Voss & Musall, 2013). Smart grids emerge as instrumental players in NZEB success, enhancing grid stability, supporting renewable energy integration, and enabling demand response programmes. Innovative concepts such as energy storage and load matching hold the promise of further elevating the sustainability of NZEBs (Deng et al., 2014).

Type of Balance

Three types of energy balance methods are used for Net Zero Energy Buildings (NZEBs), each with its characteristics:

i. **Import/Export Balance**

This method calculates the balance between exported and delivered energy, considering factors like the self-consumption of on-site generated energy. It is a comprehensive approach but can be challenging to estimate self-consumption and requires detailed simulations.

ii. **Load/Generation Balance**

This simpler approach focuses on the balance between energy generation and load within the building but doesn't consider interactions with the grid. It is easier to implement and aligns well with existing building codes.

iii. **Monthly Net Balance**

This method calculates the balance on a monthly basis, assuming that generation and load within each month cancel each other out, with only the monthly differences added up annually. It is a compromise between the other two methods, providing some insight into seasonal interactions with the grid.

Each method has its advantages and drawbacks. The import/export balance offers comprehensive information but is complex to implement. The load/generation balance is simpler but ignores grid interactions. The monthly net balance strikes a balance between simplicity and insight into seasonal energy exchanges with the grid.

The choice depends on the specific goals and available data during the design and monitoring phases of NZEBs (Sartori et al., 2012).

Temporal Energy Interplay

Evaluating the performance of Net Zero Energy Buildings (NZEBs) demands a comprehensive analysis that extends beyond annual assessments to include smaller time intervals like monthly, daily, or even hourly periods. This finer-grained evaluation is pivotal in understanding how efficiently NZEBs align their energy demand with supply over the course of a year (Deng et al., 2014). It revolves around the concept of temporal energy interplay, focusing on the building's ability to harmonize energy consumption and production over time. This synchronization is vital to strike a balance between energy utilization and generation, optimizing efficiency and the building's interaction with the local electrical grid (Sartori et al., 2012).

The concept of "***load generation alignment***" emphasizes the significance of harmonizing a building's energy generation with its energy consumption over varying timeframes, be it hourly, daily, or monthly. This alignment depends on factors like building function, the energy source, and the chosen time interval for computations. Achieving a strong correlation between load and generation contributes to efficient energy use. Conversely, an imbalance, such as excess generation during low-demand periods, leads to grid reliance. The load match index quantifies this relationship for specific energy carriers and timeframes. Improving load matching involves strategies like Demand Side Management (DSM) and Home Energy Management Systems (HEMS), which optimize energy consumption, reduce grid dependence, and promote sustainability (Deng et al., 2014; Sartori et al., 2012).

"***Building-grid interaction***" assesses the way a NZEB interacts with the utility grid, quantifying it as a ratio of net grid metering over a specific period to the maximum within a year. This metric evaluates average grid stress from factors beyond the NZEB's control (Deng et al., 2014). Monitoring energy exchanges with the grid is crucial for addressing grid challenges. Various indicators gauge grid interaction with flexible time resolutions (e.g. hourly, daily, monthly). NZEBs must be flexible to adapt energy use to grid signals. Designers play a pivotal role in optimizing this, evaluating scenarios prioritizing energy export or load matching. Access to consumption data and dynamic simulation tools is essential for crafting adaptable NZEBs (Sartori et al., 2012). Grid optimization involves not just the quantity but also the timing of electricity exports, assessed via metrics like the "fluctuation range." Grid interaction considers local infrastructure, energy sources, and "peak power demand" to determine grid capacity needs (Voss et al., 2011). The evaluation may shift towards holistic grid integration, recognizing the impact of power quality and quantity fluctuations on NZEB performance and grid stability and efficiency (Deng et al., 2014).

Energy Forecast, Measurement, and Monitoring

Achieving Net Zero Energy Building (NZEB) status necessitates a multifaceted approach that includes energy forecasting and monitoring. Energy modelling is a critical tool for estimating energy consumption, sizing on-site renewable energy sources, and identifying opportunities for efficiency improvements (Crawley et al., 2009; Feng et al., 2019). By focusing on substantial on-site renewable energy generation, the need for extensive energy efficiency measures can be reduced (Feng et al., 2019). However, this initial forecast is just the starting point in the journey toward NZEB. Robust Monitoring and Verification (M&V) processes are essential, and the choice between Measured Ratings and Calculated Ratings plays a pivotal role. Measured ratings provide real-world data from installed meters and sensors, facilitating the validation of NZEB achievement and the assessment of design effectiveness and policy compliance (Sartori et al., 2012).

Furthermore, monitoring serves a dual purpose in NZEB. It not only validates the NZEB definition but also ensures compliance with comfort standards and detects anomalies resulting from technical issues, changing conditions, or deviations from the original design (Sartori et al., 2010). Tolerance limits are established to accommodate the unpredictability of climate and occupancy variations. In essence, energy forecasting sets the initial trajectory for NZEB, while ongoing monitoring ensures that the building maintains its energy balance, temporal matching aspects, and indoor comfort (Sartori et al., 2010).

Discussion and Conclusions

The evolution of Net Zero Energy Buildings (NZEBs) reveals a fascinating interplay between global events, technological advancements, and shifting environmental awareness. From the initial wartime impetus to harnessing scarce resources, the journey unfolds through periods of ground-breaking innovation spurred by crises like the oil crisis of the 1970s. Pioneering architects like Keck and Kahn laid the groundwork with passive solar design, while landmarks like the Bliss House and Zero Energy Home solidified the concept's technical feasibility through active strategies.

Crucial milestones, such as the Passivhaus Standard and the IPCC's First Assessment Report, provided not only practical regulations but also a deeper understanding of the need for NZEBs in mitigating climate change. This led to the refinement of definitions, shifting from "zero energy houses" to a nuanced consideration of energy sources (Gilijamse, 1995). Furthermore, specific balancing metrics of source, site, cost and emission were introduced (Torcellini et al., 2006). These definitions were further improved by the introduction of Renewable Energy Supply options, categorizing them based on place of generation, ranging from on-site installations to procurement from external sources, enabling a clearer understanding of the diverse renewable energy supply options available (Pless & Torcellini, 2010).

The acknowledgement of the criticality of balancing metrics, timeframes, and system boundaries spurred the emergence of definitions grounded in metrics such as exergy (Kilkis, 2007) and lifecycle assessment (Hernandez & Kenny, 2010). Consequently, the consideration of weighted systems prompted the formulation of definitions for nearly zero and net positive energy buildings (Cole & Fedoruk, 2014), broadening the comprehension of NZEBs and necessitating a more systematic classification approach. Dialogues surrounding building-grid interaction and capabilities further refined the concept, paving the way for understanding Autonomous Net Zero Energy Buildings.

Recent NZEB classifications by Taherahmadi et al. (2021) emphasized balancing metrics, timeframes, and grid interactions but overlooked the integration of renewable energy systems, whereas Kaewunruen et al. (2019) focused solely on delineating renewable energy options and the potential for further classifying RES options.

The evolving landscape of NZEB definitions, driven by technological advancements and theoretical refinements, has exposed a critical need for a standardized evaluation framework. Such a framework would facilitate comprehensive global comparisons and enable robust classification of diverse NZEB projects. However, a key challenge lies in maintaining the contextual relevance and critical rigour of NZEB assessments within a standardized framework. In other words, while standardization fosters universal comparability, ensuring responsiveness to specific contexts (climate, cultural practices, economic realities) remains paramount. Consequently, the development of a standardized framework for NZEB evaluation must prioritize adaptability, allowing for tailoring to the unique circumstances of each project (Deng et al., 2014; Mlecnik et al., 2011; Pless & Torcellini, 2010; Sartori et al., 2012) (Fig. 3.3).

These chronicles of the chronological development of the NZEB study underscore the importance of amalgamating diverse parameters to formulate a precise definition for the building system boundary within a standardized framework. This involves considering elements such as the balancing metric and timeframe, as well as evaluating renewable supply options to delineate the net energy balance for assessing building systems. The proposed framework comprises critical components, including:

i. Building System Boundary: Defines the physical and balance boundary for assessing energy performance. This includes setting boundary conditions for functionality, space effectiveness, climate, and comfort.
ii. Weighting System: Establishes the balancing metric (e.g. primary energy, carbon emissions) and accounting period (annual, monthly) for evaluating energy consumption and generation. Symmetry across lifecycle stages (cradle-to-grave) ensures consistent accounting.
iii. Net Zero Balance Achievement: Implements energy efficiency measures to optimize indoor environmental quality while minimizing energy demand. Suitable energy supply options and grid infrastructure are identified to define the type of net zero balance pursued (on-site, grid-connected, etc.).

Fig. 3.3 Standardized framework for NZEB (*Source* Author)

iv. Temporal Energy Considerations: Accounts for the dynamic nature of energy by analysing load generation matching, building-grid interaction, and energy use/generation patterns across different timeframes (hourly, daily, seasonal).
v. Performance Assessment: Utilizes energy forecasts, real-time measurements, and ongoing monitoring to track progress towards net zero goals. This feedback loop enables continuous improvement and optimization of energy-related systems.

Establishing harmony between an internationally standardized framework and locally adaptable definitions is essential to meet the diverse requirements of different regions and contexts. The review underscores the importance of a transparent and standardized NZEB framework, such that context specific definition can be developed. This will become a fundamental platform for enhancing understanding, ensuring regulatory adherence, and facilitating the exchange of knowledge in the development of Net Zero Energy Buildings.

References

Attia, S. (2018). Net zero energy buildings (NZEB). *Elsevier*. https://doi.org/10.1016/C2016-0-03166-2

Attia, S., & Carlucci, S. (2015). Impact of different thermal comfort models on zero energy residential buildings in hot climate. *Energy and Buildings, 102*, 117–128. https://doi.org/10.1016/j.enbuild.2015.05.017

Balcomb, J. D., Hedstrom, J. C., & Perry, J. E. (1980, May 26). Performance evaluation of the Balcomb solar house. In *Annual Conference of the International Solar Energy Society*. https://www.osti.gov/biblio/6517627/#:~:text=The%20heating%20performance%20of%20the%20Balcomb%20passive%20solar,of%20data%20taken%20over%20the%20winter%20of%20197

Barber, D. (2023). *Solar architecture and solar design, 1930s–1950s—Energy history*. Yale University. https://energyhistory.yale.edu/solar-architecture-and-solar-design-1930s-1950s/

Besant, R. W., Dumont, R. S., & Schoenau, G. (1979). The Saskatchewan conservation house: Some preliminary performance results. *Energy and Buildings, 2*(2), 163–174. https://doi.org/10.1016/0378-7788(79)90031-8

Bliss, R. (1955). Design and performance of nation's only fully solar-heated house. *Air Conditioning, Heating and Ventilating, 52*(10), 92. https://api.semanticscholar.org/CorpusID:107352480

Cole, R. J., & Fedoruk, L. (2014). Shifting from net-zero to net-positive energy buildings. *Building Research & Information, 43*(1), 111–120. https://doi.org/10.1080/09613218.2014.950452

Crawley, D., Pless, S., & Torcellini, P. (2009). Getting to net zero. *ASHRAE Journal, 51*(9), 18–25. https://www.nrel.gov/docs/fy09osti/46382.pdf

Delmastro, C., & Chen, O. (2022, July 11). *Buildings-Energy System—IEA*. International Energy Agency (IEA). https://www.iea.org/energy-system/buildings

Deng, S., Wang, R. Z., & Dai, Y. J. (2014). How to evaluate performance of net zero energy building—A literature research. In *Energy* (Vol. 71, pp. 1–16). Elsevier Ltd. https://doi.org/10.1016/j.energy.2014.05.007

EPDB. (2010). *Directive 2010/31/EU of the European parliament and of the council of 19 May 2010 on the energy performance of buildings* (pp. I. 153/13–I. 153/35). The European Parliament and the Council of the European Union. https://eurlex.europa.eu/LexUriServ/LexUriServ.do?uri=OJ:L:2010:153:0013:0035:en:PDF

Esbensen, T. V., & Korsgaard, V. (1977). Dimensioning of the solar heating system in the zero energy house in Denmark. *Solar Energy, 19*(2), 195–199. https://doi.org/10.1016/0038-092X(77)90058-5

Feng, W., Zhang, Q., Ji, H., Wang, R., Zhou, N., Ye, Q., Hao, B., Li, Y., Luo, D., & Lau, S. S. Y. (2019). A review of net zero energy buildings in hot and humid climates: Experience learned from 34 case study buildings. *Renewable and Sustainable Energy Reviews, 114*, 109303. https://doi.org/10.1016/j.rser.2019.109303

Gilijamse, W. (1995). Zero-energy houses in the Netherlands. *Proceedings of Building Simulation', 95*, 276–283.

Hayes, J., & Andrejko, D. A. (1983). *Progress in passive solar energy systems. Volume 8. Part 1*. American Solar Energy Society Inc.

Healy, W. M., Gates, C., Fanney, A. H., & Pettit, B. (2015). *Design Challenges of the NIST Net Zero Energy Residential Test Facility*. https://doi.org/10.6028/NIST.TN.1847

Hernandez, P., & Kenny, P. (2010). From net energy to zero energy buildings: Defining life cycle zero energy buildings (LC-ZEB). *Energy and Buildings, 42*(6), 815–821. https://doi.org/10.1016/j.enbuild.2009.12.001

IEA. (2021). *Net zero by 2050 a roadmap for the global energy sector*

Iqbal, M. T. (2003). A feasibility study of a zero energy home in newfoundland. *Renewable Energy, 29*(2), 277–289. https://doi.org/10.1016/S0960-1481(03)00192-7

Jaysawal, R. K., Chakraborty, S., Elangovan, D., & Padmanaban, S. (2022). Concept of net zero energy buildings (NZEB)—a literature review. *Cleaner Engineering and Technology, 11*, 100582. https://doi.org/10.1016/J.CLET.2022.100582

Kaewunruen, S., Rungskunroch, P., & Welsh, J. (2019). A digital-twin evaluation of net zero energy building for existing buildings. *Sustainability (Switzerland), 11*(1). https://doi.org/10.3390/su11010159

Katzenbach, R., Clauss, F., & Zheng, J. (2014). Efficiency for sustainability—Smart energy management in social housing. *Journal of Power and Energy Engineering, 02*(12), 1–6. https://doi.org/10.4236/jpee.2014.212001

Kilkis, S. (2007). A new metric for net-zero carbon buildings. In *ASME 2007 Energy Sustainability Conference* (pp. 219–224). https://doi.org/10.1115/ES2007-36263

Kurnitski, J., Allard, F., Heiselberg, P., & Kosonen, R. (2011). *How to define nearly net zero energy buildings nZEB-REHVA proposal for uniformed national implementation of EPBD recast OpenBPS*. https://www.researchgate.net/publication/284670277

Levy, B. (2017, April 25). *The first U.S. house to go solar*. MIT Technology Review. https://www.technologyreview.com/2017/04/25/152254/the-first-us-house-to-go-solar/

Lewis, N. S. (2007). Toward cost-effective solar energy use. *Science, 315*(5813), 798–801. https://doi.org/10.1126/SCIENCE.1137014

Madathil, D., Pandi, V. R., Nair, M. G., Jamasb, T., & Thakur, T. (2021). Net zero energy in a residential building using heuristic optimization solution. *Journal of Control, Automation and Electrical Systems, 32*(2), 458–471. https://doi.org/10.1007/s40313-020-00685-7

Marszal, A.J., & Heiselberg, P. (2009). *A literature review of zero energy buildings (ZEB) definitions*. https://vbn.aau.dk/ws/portalfiles/portal/18915080/A_Literature_Review_of_Zero_Energy_Buildings__ZEB__Definitions

Marszal, A. J., Heiselberg, P., Bourrelle, J. S., Musall, E., Voss, K., Sartori, I., & Napolitano, A. (2011). Zero energy building—A review of definitions and calculation methodologies. *Energy and Buildings, 43*(4), 971–979. https://doi.org/10.1016/j.enbuild.2010.12.022

Mlecnik, E., Attia, S., & Van Loon, S. (2011, October). Net zero energy building: a review of current definitions and definition development in Belgium. *Proceedings of Passive House*. www.passiefhuisplatform.be

Moghaddasi, H., Culp, C., Vanegas, J., & Ehsani, M. (2021). Net zero energy buildings: Variations, clarifications, and requirements in response to the Paris agreement. *Energies, 14*(13), 3760. https://doi.org/10.3390/en14133760

Omrany, H., Chang, R., Soebarto, V., Zhang, Y., Ghaffarianhoseini, A., & Zuo, J. (2022). A bibliometric review of net zero energy building research 1995–2022. *Energy and Buildings, 262*, 111996. https://doi.org/10.1016/j.enbuild.2022.111996

Pless, S., & Torcellini, P. (2010). *Net-Zero Energy Buildings: A Classification System Based on Renewable Energy Supply Options*. https://doi.org/10.2172/983417

Saini, L., Meena, C. S., Raj, B. P., Agarwal, N., & Kumar, A. (2022). Net zero energy consumption building in India: An overview and initiative toward sustainable future. *International Journal of Green Energy, 19*(5), 544–561. https://doi.org/10.1080/15435075.2021.1948417

Saitoh, T. (1984). Natural energy autonomous house with underground water reservoir. *Transactions of the Japan Society of Mechanical Engineers. B, 50*(452), 1044–1044. https://doi.org/10.1299/KIKAIB.50.1044

Sartori, I., Napolitano, A., Marszal, A., Pless, S., Torcellini, P., & Voss, K. (2010). Criteria for definition of net zero energy buildings. *Proceedings of the EuroSun 2010 Conference*, 1–8. https://doi.org/10.18086/eurosun.2010.06.21

Sartori, I., Napolitano, A., & Voss, K. (2012). Net zero energy buildings: A consistent definition framework. *Energy and Buildings, 48*, 220–232. https://doi.org/10.1016/j.enbuild.2012.01.032

Shetty, A., Samarth, K. S., Naik, S. K., Setty, H. T. M., & Hannah, K. (2023). *Retrofit guidelines towards achievement of net zero energy in a university building* (pp. 231–243). https://doi.org/10.1007/978-981-19-1862-9_15

Shirinbakhsh, M., & Harvey, L. D. D. (2021). Net-zero energy buildings: The influence of definition on greenhouse gas emissions. *Energy and Buildings, 247*. https://doi.org/10.1016/j.enbuild.2021.111118

Singh, B., Sharma, S. K., & Syal, P. (2021). Net zero energy building (NZEB) design. *Journal of the Institution of Engineers (India): Series A, 102*(1), 237–247. https://doi.org/10.1007/s40030-020-00500-1

Stahl, W., Voss, K., & Goetzberger, A. (1994). The self-sufficient solar house in Freiburg. *Solar Energy, 52*(1), 111–125.

Taherahmadi, J., Noorollahi, Y., & Panahi, M. (2021). Toward comprehensive zero energy building definitions: A literature review and recommendations. *International Journal of Sustainable Energy, 40*(2), 120–148. https://doi.org/10.1080/14786451.2020.1796664

Torcellini, P., Pless, S., Deru, M., & Crawley, D. (2006). Zero energy buildings: A critical look at the definition. *2006 ACEEE Study on Energy Efficiency in Buildings*, 3/417–3/428. https://www.osti.gov/biblio/883663

Voss, K., & Musall, E. (2013). Net zero energy Buildings. In C. Hellstern, K. Ahrend, & J. Rackwitz (Eds.), *Green books* (Vol. 58, Issue 7). Institut für internationale Architektur-Dokumentation. http://dnb.d-nb.de

Voss, K., Musall, E., & Lichtmeß, M. (2011). From low-energy to net zero-energy buildings: Status and perspectives. *Journal of Green Building, 6*(1), 46–57. https://doi.org/10.3992/JGB.6.1.46

Voss, K., Sartori, I., Napolitano, A., Geier, S., Gonçalves, H., Hall, M., Heiselberg, P., Widén, J., Candanedo, J. A., Musall, E., Karlsson, B., & Torcellini, P. (2010). Load matching and grid interaction of net zero energy buildings. In *Proceedings of the EuroSun 2010 Conference* (pp. 1–8). https://doi.org/10.18086/eurosun.2010.06.24

Wu, W., & Skye, H. M. (2021). Residential net-zero energy buildings: Review and perspective. In *Renewable and Sustainable Energy Reviews* (Vol. 142). Elsevier Ltd. https://doi.org/10.1016/j.rser.2021.110859

Chapter 4
Water Availability and Higher Temperatures Affecting the Behavior, The Chemical Composition, and the Antimicrobial Activities of *Senna italica* Plants: A Study in Fujairah, UAE

François Mitterand Tsombou, Aisha Ahmed Mohamed Al Kaabi, Aishah Saeed Sulaiman Al Dhanhani, Shaher Bano, and Fouad Lamghari Ridouane

Abstract Water availability and higher temperatures are the major ecological factors affecting plant metabolism in arid systems. The current study explores the combined effects of water availability at three levels (6, 12, and 18%) and higher temperatures on the plant attributes of *Senna italica*. To understand this, we grew *S. italica* plants in the net house and under natural conditions during the hot summer of Fujairah. According to our results, the root length of plants growing under field conditions was positively affected by the drought and temperatures. Contrary, more biomasses were allocated to the shoots of those growing in the net house. Survival rates were significantly lower for plants growing in the field. Minerals assessments revealed that plants growing in the field had greater values of calcium, magnesium, and potassium, whereas those of the net house showed more phosphorus, zinc, tannin, ash, carotenoids, and flavonoids. We found higher zones of inhibition 14.33, and 14 mm against *E. coli*, *S. aureus*, and *C. albicans* for plant root growing under natural conditions. The present study indicates that *S. italica* could be a model plant for land degradation restoration in the

---

F. M. Tsombou (✉) · A. A. M. Al Kaabi · A. S. S. Al Dhanhani · S. Bano · F. L. Ridouane
Fujairah Research Centre, Fujairah, UAE
e-mail: francois.tsombou@frc.ae

A. A. M. Al Kaabi
e-mail: aisha.alkaabi@frc.ae

A. S. S. Al Dhanhani
e-mail: aishah.suliman@frc.ae

S. Bano
e-mail: shaher.bano@frc.ae

F. L. Ridouane
e-mail: fouad.lamghari@frc.ae

© The Author(s), under exclusive license to Springer Nature Switzerland AG 2026
J. Dsilva (ed.), *Redefining Sustainability*, https://doi.org/10.1007/978-3-031-88376-7_4

desert regions. This study also indicates that plant metabolites resulting from higher drought stress might be beneficial for pharmaceutical uses.

Keywords Drought · Higher temperatures · Plant-*Senna italica* · Water availability

Introduction

Water availability and higher temperatures during the summer in the driest places are the major environmental factors that significantly alter plant physiology (Ayanlade et al., 2022; Shi et al., 2022). To understand how these two factors impact on different plant attributes, then the current study attempts to understand how water availability and higher temperatures affect plant growth and antimicrobial activities of *Senna italica* during the hottest times in the Emirate of Fujairah in the net house and natural conditions.

Recent studies on climate changes revealed that temperatures have raised up to 1.5 °C (Pringle, 2018), and this is expected to reach greater levels in the coming years. The impacts of water deficiencies on plants physiology and crop production are disastrous (Farooq et al., 2012; Xiong et al., 2006). Drought stress showed adverse effects on plant nutritional components and mineral composition (Lashley & Harper, 2012; Madouh, 2022).

Many earlier studies argued that drought stress induces reactive oxygen species (ROS) in plant facing drought stress (Cruz, 2008; Du et al., 2020; Mittler & Blumwald, 2015; Sachdev et al., 2021). Mineral and phytochemical composition adjustments and enzymatic processes are the main key factors that allow plants to survive under water deficit. With this regard, calcium, potassium, and zinc concentrations were seen to increase significantly under water deficiencies (Hassan et al., 2020; Jing et al., 2019; Raza et al., 2013); while the levels of phosphorous and magnesium were greatly reduced (Santos et al., 2023; Smith et al., 2022).

mPlants of the desert regions have significantly adjusted their ways to cope with embarrassing ecological issues. *Senna italica* is considered one of the desert plants with great potentialities to tolerate long-term drought. This plant species presents significant uses for the local people of UAE. Plants of those milieus face more stress than the others, and therefore, novel metabolites could be synthetized while facing such ecological issues. Those bioactive components might have interesting benefits against some higher resistant pathogens such as *Staphylococcus aureus*, *Candida albicans*, and *Escherichia coli*. Unfortunately, the literature review does not provide enough evidential information related to plant drought stress and antimicrobial activities, and nutritional contents in the arid and semi-arid regions generally.

Our findings revealed that water limitations strongly affect plants' growth of *Senna italica* in the summer. Furthermore, we found that plants facing more drought stress had greater antimicrobial activities than the others.

Methodology

Site Study and Importance of the Work

The current study was carried out in the Emirates of Fujairah, United Arab Emirates (UAE) (25.1288° N, 56.3265° E) from May to July 2023. The experimental months are known to be hottest in UAE. *Senna italica* is one of the most important perennial plants in UAE. The Emirate of Fujairah like the other Emirates is involved in many afforestation projects with the main goal to re-green the city. Unfortunately, those projects are delicate for implementation due to the harsh conditions, especially during the summer, as the introduced plants have very little chance of surviving. Furthermore, plants of the extreme environments face more stress than the other and novel metabolites could be synthesized while facing those embracing factors. Those metabolites could help fight against some virulent pathogens related to human health.

Seeds and Soil Samples Collection

Mature fruits of *Senna italica* were collected from the natural habitat and seeds were extracted, and then stored in the paper bags at room temperature (20 ± 2 °C) for further use. Desert sand soil was also collected at the depth 0–20 cm, air-dried, and then passed through a 2 mm sieve.

Experimental Design

Free draining plastic pots (23-cm diameter) were filled with 5 kg dried soils (60% sands + 40% potting soils). The experiment was carried out with 15 replicates per group, and three levels of soil moisture such as 6, 12, and 18%. Thereafter, we placed 25 seeds of *Senna italica* in each pot, and all the pots were kept in the net house to allow the seeds to germinate. After germination, half of the pots were moved in the field, and acclimatized for one week. Then, water treatments were applied, and the watering was scheduled for two times per week. After 75 days, the experimented plants were harvested and the following parameters were measured: shoot length, root length, shoot dry weight, root dry weight, number of leaves, mortality rate, flavonoids, carotenoids, ash, calcium, magnesium, phosphorous, zinc, potassium, tannin, antimicrobial activities, and environmental factors. Atmospheric temperature and the relative humidity were recorded daily and three times per day (early morning, noon time, and evening) using the hydrometer.

Nutrients Determination

The extractions of potassium and sodium were done according to Association of Official Agricultural Chemists (AOAC 969.23). Calcium, magnesium, phosphorous, and zinc were extracted following (AOAC 2015.01). Tannin extraction was performed by titrating the extract with the standard potassium permanganate solution through indigo-carmine solution and KMnO4 solution to obtain the light yellowish color end point. Ash content was performed following (AOAC 942.05).

Flavonoids and Carotenoids Determinations

Flavonoids contents extraction was done by mincing plant part and macerating with methanol (1:20, w/v) for 72 h at room temperature (28 ± 2 °C). The extract was then filtered, and the obtained residue was re-macerated with the same solvent until the extraction was exhausted. The filtrate obtained from the maceration results was subsequently combined, and then evaporated with a rotary evaporator. Afterward, samples were incubated for 30 min at room temperature, and then the absorbance was measured on UV–Vis spectrophotometry at a wavelength of 431 nm. The readings were done from three replications for each analysis. Carotenoid extraction was done from the stock solution of 20 000 μg/100 ml) and working solution of beta carotene. Then 25 g of the powder of each sample was diluted in 200 ml at 35 °C. Approximately 5.25 g reconstituted sample was transferred into a 50 ml centrifuge tube. 5 ml of the appropriate internal standard solution was pipetted, and 1.5 ml KOH was added to each tube and shooked for 5 min. A reconstituted sample solution was injected to UHPLC, the concentration of beta carotene was obtained from the calibrated curved.

Antimicrobial Activities

Staphylococcus aureus and *Escherichia coli* were the bacteria strains, while *Candida albicans* was the fungal strain. Antimicrobial activities of *Senna italica* plant extracts were determined on Mueller Hinton agar using agar disc method (Hudzicki, 2016).

Statistical Analysis

Two-way ANOVAs were performed to test the effects of water availability and plant location, and plant parts on the growth parameters, chemical composition, and the antimicrobial activities. All the data were performed with SYSTAT 13.

Results

Environmental Parameters

The averages temperatures in the net house were 32.84, 43.94, and 36. 52 °C, respectively, for the morning, noon time, and evening; and those of the humidity were 70.94, 37.48, and 55.74% for similar periods. In the field, the values of the temperatures were 34.20, 45.58, and 35.81 °C, respectively, for the morning, noon time, and evening; and those of the humidity were 70.82, 32.4, and 54.57%.

Impacts of Water Availability and Plant Location on the Growth of Senna italica

Water availability and plant location had significant effects on the morphological parameters of *Senna italica* (Fig. 4.1). Plant shoots growing in the field at 12% were strongly reduced (3.4 cm) compared than 18% (7.33 cm). Plant root growing in the field showed higher length at 12% of soil moisture (47.6 cm) than that of 18% (44.66 cm). Plant leaves showed important growth in the net house at 18% (26.11) than 12 and 6%. Shoot biomasses of the plant growing at 12% of soil moisture in the field were adversely affected (1.14 g) than that of 18% (1.4 g). Plant growing in the net house at 18% showed higher root dry biomass (2.7 g) than 12 and 6%. Less root dry weight (0.36 g) was recorded at 12% for the plant growing in the field. Plant growing at 6% (100%) of soil moisture under field conditions did not survive compared to 12% (66.66%) and 18% (17.03%).

Effects of Plant Parts and Drought Chemical Composition of Senna italica Plants

Plant parts and drought had significant effects on the chemical composition of *Senna italica* plants (Fig. 4.2a, b). Plant growing in the field under 18% of soil moisture showed more important calcium (776. 84 mg/100 g), magnesium (287.58 mg/100 g), and potassium (920.27 mg/100 g), respectively, for the roots and shoots. The levels of phosphorous (100.47 mg/100 g), zinc (1.67 mg/100 g), tannin (1.21%), and ash (3.215%) were greater in the plant shoots and roots growing at 6% of soil moisture in the net house. Plant shoots growing in the net house at 18% revealed greater contents in carotenoids (13,488 mg/kg), and flavonoids (1624 mg/kg) than those of 12 and 6%. The levels of carotenoids and flavonoids were shoot > root, net house > field, and 18% > 12% > 6%.

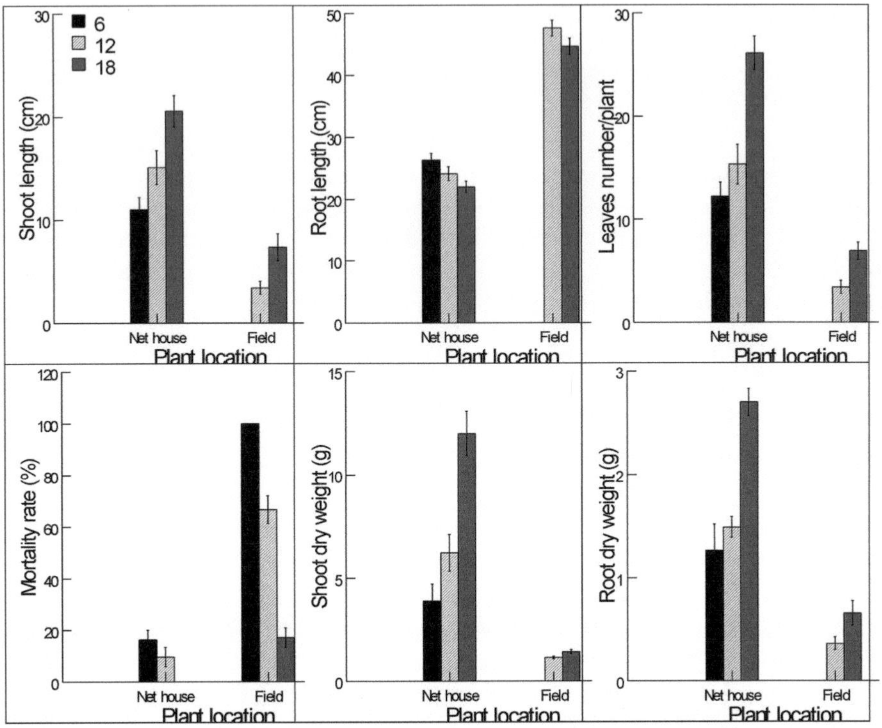

Fig. 4.1 Effects of water stress (6, 12, and 18%) and the plant location (net house and field) on the plant growth of *Senna italica*

Effects of Plant Parts of Senna italica and Drought on the Antimicrobial Activities of Staphylococcus aureus, Candida albicans, and Escherichia coli

Plants parts had no significant effects on the microbial activities of *Staphylococcus aureus*, *Candida albicans*, and *Escherichia coli* (Fig. 4.3a, b). Water availability revealed significant effects on the harvested plant and their microbial activities against the tested pathogens. Plant roots growing in the field showed higher antimicrobial effects than those of the net house, and bacteria were more affected than the fungus. The inhibition zones were (14.33 and 14 mm) against *E. coli*, *S. aureus*, and *C. albicans*.

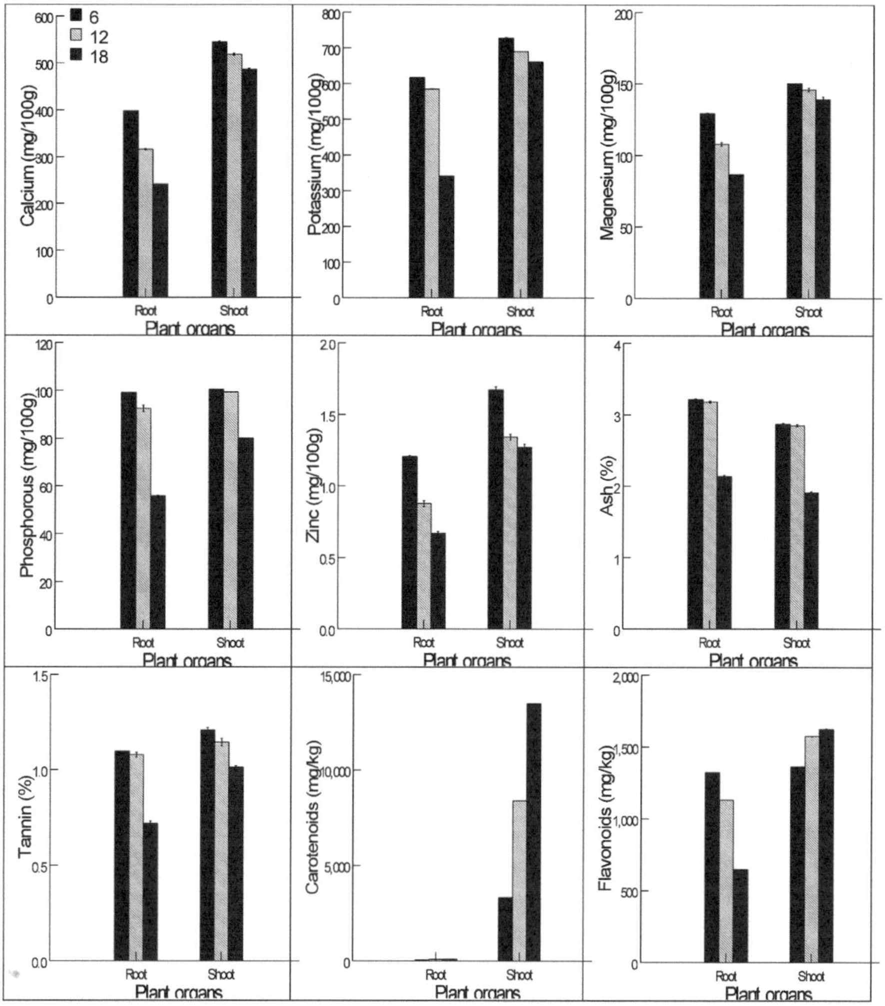

Fig. 4.2 Nutrients composition of the shoot and root of *S. italica* growing under three levels of drought in the net house (**a**) and field conditions (**b**)

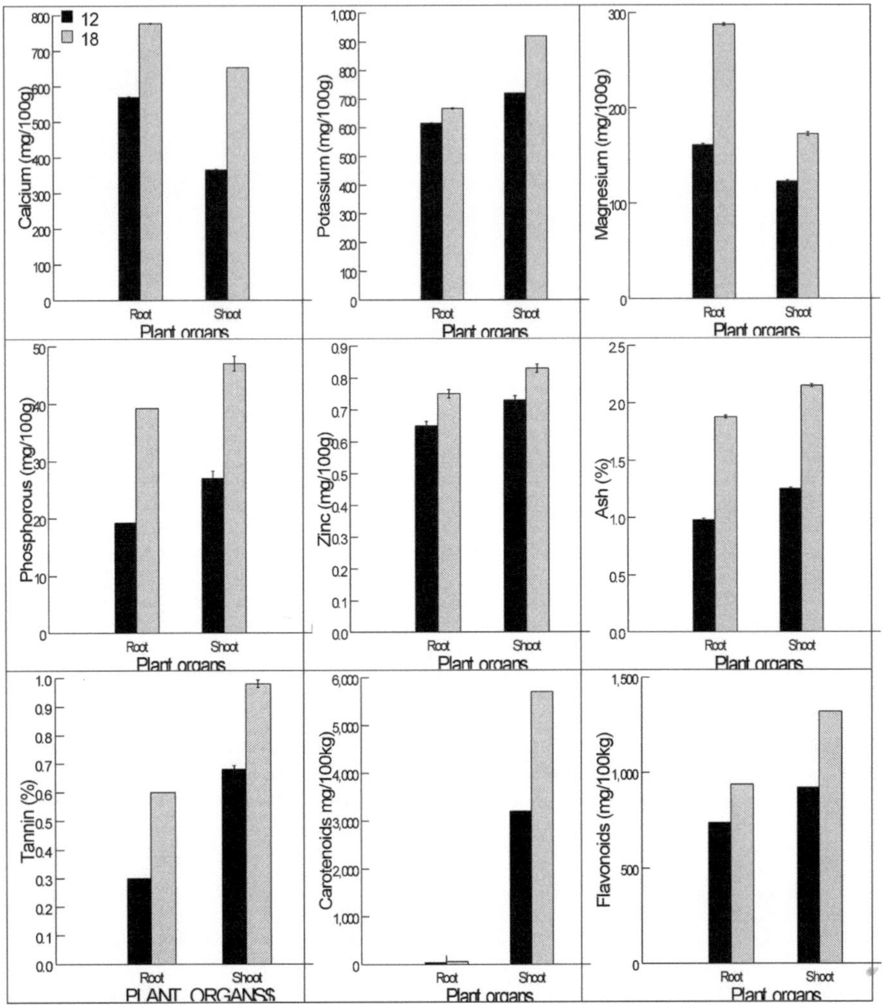

Fig. 4.2 (continued)

Conclusion

Our findings revealed that mineral balance between plant shoot and root is very important for plants growing under extreme environments. We also demonstrated that plant growing under water limitations present more important antimicrobial activities. The current findings proved that the tested plant can be a model plant for land restoration in arid ecosystems. Moreover, deep analyses on the phytochemical composition could allow to identify metabolites with significant effects on the microbial activities under hottest times of the year.

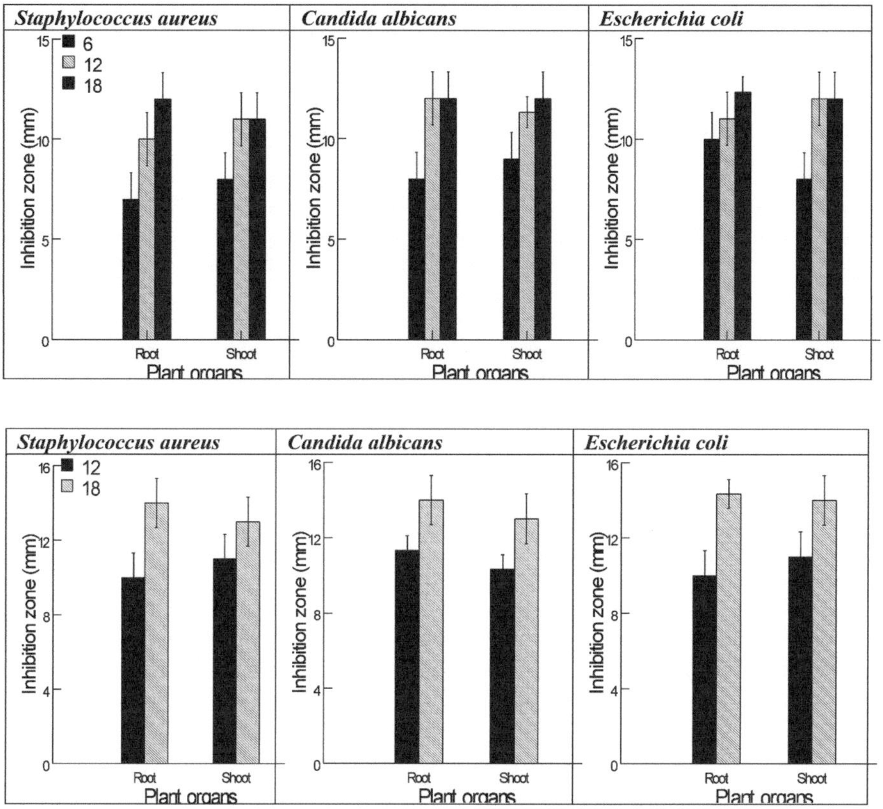

Fig. 4.3 Antimicrobial activities of *Senna italica* plants and drought growing in the net house (**a**) and field (**b**) against *Staphylococcus aureus*, *Candida albicans*, and *Escherichia coli*

References

Ayanlade, A., Oladimeji, A., Okegbola, O., Eludoyin, A., Eslamian, S., Ayinde, A., & Perkins, P. (2022). Effect of climate change on water availability and quality: An assessment of socio-resilience in Nigeria. In *Disaster risk reduction for resilience* (pp. 245–262).

Cruz, D. C. M. H. (2008). Drought stress and reactive oxygen species: Production, scavenging and signaling. *Plant Signaling & Behavior, 3*(3), 156–165.

Du, Y., Zhao, Q., Chen, L., Yao, X., & Xie, F. (2020). Effect of drought stress at reproductive stages on growth and nitrogen metabolism in soybean. *Agronomy, 10*, 1–21.

Farooq, M., Hussain, M., Wahid, A., & Siddique, K. (2012). Drought stress in plants: An overview. In *Plant responses to drought stress* (pp. 1–33).

Hassan, M. U., Aamer, M., Chattha, M. U., Haiying, T., Shahzad, B., Barbanti, L., Nawaz, M., Rasheed, A., Afzal, A., Liu, Y., & Guoqin, H. (2020). The critical role of zinc in plants facing the drought stress. *Agriculture, 10*, 1–20.

Hudzicki, J. (2016). Kirby-Bauer disk diffusion susceptibility test protocol. American Society for Microbiology.

Jing, X., Cai, C., Fan, S., Wang, L., & Zeng, X. (2019). Spatial and temporal calcium signaling and its physiological effects in moso bamboo under drought stress. *Forest, 10*, 1–15.

Lashley, M. A., & Harper, C. A. (2012). The Effects of extreme drought on native forage nutritional quality and white-tailed deer diet selection. *Southeastern Naturalist, 11*(4), 699–710.

Madouh, T. A. (2022). Eco-physiology responses of native desert plant species to drought and nutritional levels: Case of Kuwait. *Frontiers in Environmental Science, 10*, 1–13.

Mittler, R., & Blumwald, E. (2015). The roles of ROS and ABA in systemic acquired acclimation. *The Plant Cell, 27*, 64–70.

Pringle, P. (2018). Effects of climate change on 1.5° temperature rise relevant to the Pacific islands. In *Pacific marine climate change report card, science review* (pp. 189–200).

Raza, M. A. S., Saleem, M. F., Shah, G. M., Jamil, M., & Khan, I. H. (2013). Potassium applied under drought improves physiological and nutrient uptake performances of wheat (*Triticum Aestivun* L.). *Journal of Soil Science and Plant Nutrition, 13*(1), 175–185.

Sachdev, S., Ansari, S. A., Ansari, M. I., Fujita, M., & Hasanuzzaman, M. (2021). Abiotic stress and reactive oxygen species: Generation, signaling, and defense mechanisms. *Antioxidants, 10*, 277.

Santos, A. S., Pinho, D. S., Silva, A. C. d., Brito, R. R. D., Lacerda, J. J. D. J., Silva, E. M. D., et al. (2023). Magnesium supplementation alleviates drought damage during vegetative stage of soybean plants. *PLoS ONE, 18*(11), 1–15.

Shi, Y., Zhao, L., Zhao, X., Lan, H., & Teng, H. (2022). The integrated impact of drought on crop yield and farmers' livelihood in semi-arid rural areas in China. *Land, 11*, 2260.

Smith, M. R., Dinglasan, E., Veneklaas, E., Polania, J., Rao, I. M., Beebe, S. E., & Merchant, A. (2022). Effect of drought and low phosphorous on yield and nutritional content in common bean. *Frontiers in Plant Science* 1–14.

Xiong, L., Wang, R.-G., Mao, G., & Koczan, J. M. (2006). Identification of drought tolerance determinants by genetic analysis of root response to drought stress and abscisic acid. *Plant Physiology, 142*, 1065–1074.

Chapter 5
Exploring Various Dimensions of Adopting Renewable Solar Energy in the UAE: A Manager's Perspective

Moustafa Nour El Din Hussein Sherif

Abstract This study seeks to investigate the various aspects of integrating solar energy as a renewable source in the United Arab Emirates (UAE) through the lens of professionals in the renewable energy field. The research explores the economic, environmental, and social dimensions of incorporating solar energy, while assessing its impact on sustainable development and the influence of regulatory policies. Employing a qualitative approach grounded in exploratory and descriptive methods based on an interpretivist research philosophy, and following an inductive approach, the study employs qualitative research techniques, specifically in-depth interviews with fifteen renewable energy sector managers, to draw well-founded conclusions. The interviews revealed key themes such as climate change, sustainability, renewable energy sources, factors influencing solar energy adoption, challenges, limitations, and government initiatives and regulations. The findings highlight the pressing need to address climate change and the potential of solar energy, while emphasizing the need to address challenges such as low tariff rates and limited regulations. The study recommends addressing governance issues, promoting trade openness, collaborating with the private sector, raising awareness, investing in infrastructure, and developing national incentive programs.

Keywords Renewable solar energy · Transition toward sustainability · Qualitative research · Climate change · Green economy

Introduction

Scope and Objective of the Paper and Problem Statement:

The aim of this study is to examine different aspects of adopting renewable solar energy in the UAE from the perspective of managers. Renewable energy, particularly solar energy, has garnered significant attention due to its potential to provide

M. N. E. D. H. Sherif (✉)
Dubai, UAE
e-mail: moustafa_nour_2020@hotmail.com

© The Author(s), under exclusive license to Springer Nature Switzerland AG 2026

J. Dsilva (ed.), *Redefining Sustainability*, https://doi.org/10.1007/978-3-031-88376-7_5

an unlimited and environmentally friendly energy supply, offering remarkable environmental benefits compared to conventional energy sources (Shahsavari & Akbari, 2018). The UAE serves as a noteworthy case study for solar energy development due to its significant potential in this field, supported by clear regulations and unwavering government support, creating an ideal environment for industry growth and advancement.

Global attention toward utilizing renewable energy sources has increased significantly, with the GCC countries showing growing interest in renewable energy development (Bhutto et al., 2014). In the UAE, the solar industry is witnessing consistent growth, primarily driven by declining costs of solar systems, government policies, tax credits, and incentives (Karmakar & Sahib, 2017). However, the UAE is currently confronted with the dual challenges of climate change and the depletion of fossil fuels, necessitating a thorough exploration of alternative energy sources (Jamil et al., 2015). This study holds substantial value as it contributes to the existing literature on sustainable energy practices by exploring economic, environmental, and social aspects. Additionally, it evaluates the impacts of solar energy technology on sustainable development and the influence of policies and regulations, providing valuable insights for policymakers, industry professionals, and stakeholders to develop effective strategies for the successful implementation of renewable solar energy in the UAE, thus contributing to the attainment of sustainable development goals. The research has the following key objectives:

To conduct interviews with fifteen managers working in the renewable field to identify their perspectives on the dimensions of adopting renewable solar energy in the UAE.

- To examine the impact of solar energy technology on sustainable development in the UAE.
- To evaluate the influence of policies and regulations for adopting solar energy in the UAE.
- To recommend strategies to effectively adopt renewable solar energy in the UAE.

The integration of solar energy in the UAE has garnered significant scholarly attention, prompting a comprehensive review of relevant theories, concepts, and empirical research. The theoretical and conceptual review encompasses the Social Representations Theory (SRT) and the Innovation Diffusion Theory. SRT emphasizes the impact of macro-level social changes and the shaping of attitudes toward new energy technologies through shared social representations. It underscores the significance of relatability, perceived benefits, and addressing local resistance in the adoption of renewable energy technologies (Boudet, 2019). Innovation Diffusion Theory, developed by Rogers, focuses on the systematic spread of innovative ideas and technologies within a social system, identifying fundamental factors influencing the diffusion process and the pivotal role of different types of adopters (Wani & Ali, 2015). By integrating both theories, a holistic framework is created to comprehend the social, cultural, and managerial factors that impact the adoption of renewable energy in the UAE.

Several empirical studies have explored renewable energy deployment in the Middle East and GCC countries. Sim (2023) investigates the potential consequences of integrating renewable energy into the governance structures of the UAE, highlighting the limited immediate impact in the short term and the positive long-term outlook. Alharbi and Csala (2020) delve into the reliance of GCC countries on fossil fuels and their pursuit of renewable energy technologies, emphasizing the importance of expediting renewable energy projects. Bellakhal et al. (2019) emphasize the influence of governance and trade openness on investment in renewable energy within the MENA region. Additionally, Ferroukhi et al. (2013) evaluate the efforts and drivers of renewable energy deployment in the GCC countries, while Griffiths and Mills (2016) examine the potential of solar rooftop in the UAE and its economic viability.

Results and Findings

The primary focus of this study revolved around the following research question: How do renewable energy professionals perceive the importance of adopting renewable solar energy in the UAE?

By utilizing the SRT, the perception of renewable energy professionals regarding the significance of embracing solar energy in the UAE can be understood within the framework of large-scale social transformation processes. SRT suggests that attitudes are shaped by social representations, which encompass attitudes, beliefs, and practices rather than solely focusing on cognitive beliefs. For instance, solar energy is recognized as a major sector in the UAE, supported by existing literature highlighting its economic viability. Factors influencing solar energy adoption include grid stability, ROI, net metering regulations, sustainability, cost savings, government incentives, and economic considerations. Key value propositions identified are green energy, sustainability, energy security, and financial savings. However, barriers such as low tariff rates, lack of regulations, and high costs need to be addressed. Socio-cultural factors like lack of awareness and education also influence adoption. Government initiatives, collaboration with the private sector, and supportive policies are important, but bureaucratic processes and resistance may pose challenges. By considering the discourse and communication within and between individuals, groups, and society, the implementation of Solar Energy in the UAE can be prioritized through policy support, awareness initiatives, education, and collaboration to tackle climate change and foster a sustainable future.

Methodology

The methodology for this research is grounded in a comprehensive understanding of the research strategy, encompassing principles and philosophical beliefs, as highlighted by Melnikovas (2018). The study employs a qualitative research approach, aligning with interpretivism to capture the complexities and individual experiences associated with the adoption of solar energy in the UAE (Goldkuhl, 2012). Furthermore, the research utilizes an inductive approach, which allows for the collection and analysis of data to generate new theories or insights based on observed patterns and themes (Pascaris et al., 2021). The data collection method involves qualitative analysis, enabling a deep exploration of the complexities and individual experiences, capturing the unique dynamics of solar energy adoption in the UAE (Sandelowski, 2000). Additionally, the research employs purposive sampling to select fifteen managers from engineering organizations operating within the renewable energy sector in the UAE, aiming to gain valuable insights into the deployment of solar energy (Saunders et al., 2007). The study adopts a cross-sectional time horizon due to time constraints, focusing on assessing a specific problem within a particular timeframe (Levin, 2006). Data analysis will be conducted using the Braun and Clarke methodology to ensure accuracy and consistency of the findings (Braun & Clarke, 2006). This methodology is well-suited for exploring the dimensions of adopting renewable solar energy in the UAE, as it provides a holistic understanding of managers' viewpoints and experiences in the industry.

Findings

Main themes	Description
Climate change and sustainability	This theme focuses on the issues of global warming, renewable energy, and the urgency for sustainable solutions
Renewable energy sources and project models	This theme explores a wide range of renewable energy sources, including solar, wind, geothermal, hydrogen, biomass energy, and storage technologies
Factors influencing Solar energy adoption and value Propositions	This theme examines the factors that influence the adoption of solar energy, including the suitability and advantages of on-grid PV systems, net metering regulations, and economic factors. Additionally, it focuses on the key value propositions of solar energy, potential cost savings, enhanced energy security, and the reduction of carbon footprint

(continued)

(continued)

Main themes	Description
Challenges, limitations, pros, and cons of solar energy adoption	This theme discusses the challenges and limitations, including factors such as lack of awareness, education, cultural attitudes, and financial incentives. Additionally, it highlights the environmental benefits, potential cost savings
Factors affecting solar industry growth	This theme covers various factors that influence the growth of the solar industry, including regulations, taxation, government policies, awareness, and manufacturing costs
Government initiatives and regulations	This theme focuses on national incentive programs, net metering regulations, government investment, and collaboration with the private sector

The research findings underscore the pressing need for sustainable solutions in the face of global warming, emphasizing the UAE's strong commitment to sustainability and ecosystem protection. It delves into various renewable energy sources and project models, thoroughly examining factors influencing the adoption of solar energy, addressing challenges, and highlighting advantages such as brand enhancement and reducing reliance on oil revenue. It stresses the importance of supportive government policies, awareness programs, incentives, and private sector collaboration. The main themes encompass diverse topics illustrated in the below table. Each of these themes represents a distinct aspect relating to solar energy, and their definitions and names have undergone refinement processes to uncover their fundamental essence.

Conclusions

This research has offered valuable insights into the perspectives of managers regarding climate change, sustainability, and renewable energy adoption in the UAE. The findings emphasize the urgency of addressing climate change and transitioning to clean and sustainable energy solutions. Solar energy emerged as a major focus, reflecting its vast potential in the region. The study explored different business models and identified factors influencing solar energy adoption, such as grid stability, return on investment, and government incentives. Key value propositions for adopting renewable solar energy include being eco-friendly, enhancing energy security, and realizing financial savings. This research contributes to the understanding of business models, policy frameworks, and strategies for widespread solar energy adoption in the UAE, aligning with the country's vision for sustainable development. Addressing challenges, promoting awareness, and investing in infrastructure are recommended for further progress in this area.

Recommendations

To facilitate widespread adoption of renewable solar energy in the UAE, this study strongly recommends the following key actions:

Address Governance Issues

- Promote Trade Openness and Policy Frameworks: This involves advocating for policies that support renewable energy projects and incentivize their development.
- Collaboration with the Private Sector: Encouraging collaboration between the public sector and private companies is important for driving innovation and investment in renewable energy.

Raise Awareness

- Educate individuals and communities about the environmental and financial benefits of solar energy.
- Overcome barriers such as low tariff rates and lack of awareness.

Invest in Infrastructure

- Enhance production capacity and reliability by expanding the use of microgrid solutions.
- Prioritize the integration of solar energy into the existing grid infrastructure.

Develop National Incentive Programs

- Facilitate Net Metering Regulations: Net metering allows individuals producing their own solar energy to inject surplus energy into the grid and receive credits for it.
- Providing Financial Incentives: This can include favorable tariff rates, which means that individuals who generate solar energy can sell it back to the grid at higher rates.

Limitations of the Study

The research conducted in this study had several constraints that should be acknowledged. It focused only on the perspectives of renewable energy professionals and managers in the UAE, potentially limiting the representation of other stakeholders' opinions. The research relied on self-reported data, which could introduce response bias or inaccuracies, and it was specific to the UAE, potentially limiting its applicability to other regions due to cultural, economic, and policy differences. Future research should aim to include a broader range of participants, consider objective measurements, and carefully consider the study's findings when applying them to different geographical areas.

Future Implications

The study emphasizes the need for further investigation based on its exploratory nature. It acknowledges the limitations of a small sample size and suggests future research should involve a larger and more diverse group of participants, including personnel from the product and business sides. It also highlights the potential for bias and recommends incorporating quantitative analysis for more objective results. Additionally, it stresses the importance of investigating the long-term effects and sustainability of adopting renewable solar energy in the UAE, including assessing economic, social, and environmental impacts, identifying potential barriers, and developing effective implementation strategies.

References

Alharbi, F. R., & Csala, D. (2020). GCC countries' renewable energy penetration and the progress of their energy sector projects. *IEEE Access, 8*, 211986–212002.

Bellakhal, R., Kheder, S. B., & Haffoudhi, H. (2019). Governance and renewable energy investment in MENA countries: how does trade matter? *Energy Economics, 84*, 104541.

Bhutto, A., Bazmi, A. A., Zahedi, G., & Klemeš, J. J. (2014). A review of progress in renewable energy implementation in the Gulf Cooperation Council countries. *Journal of Cleaner Production, 71*, 168–180.

Boudet, H. S. (2019). Public perceptions of and responses to new energy technologies. *Nature Energy, 4*(6), 446–455.

Braun, V., & Clarke, V. (2006). Using thematic analysis in psychology. *Qualitative Research in Psychology, 3*(2), 77–101.

Ferroukhi, R., Ghazal-Aswad, N., Androulaki, S., Hawila, D., & Mezher, T. (2013). Renewable energy in the GCC: Status and challenges. *International Journal of Energy Sector Management, 7*(1), 84–112.

Goldkuhl, G. (2012). Pragmatism vs interpretivism in qualitative information systems research. *European Journal of Information Systems, 21*(2), 135–146.

Griffiths, S., & Mills, R. (2016). Potential of rooftop solar photovoltaics in the energy system evolution of the United Arab Emirates. *Energy Strategy Reviews, 9*, 1–7.

Jamil, M., Ahmad, F., & Jeon, Y. J. (2015). Renewable energy technologies adopted by the UAE: Prospects and challenges—A comprehensive overview. *Renewable and Sustainable Energy Reviews, 55*(1), 1181–1194.

Karmakar, A., & Sahib, U. (2017). Smart Dubai: Accelerating innovation and leapfrogging E-democracy. In T. V. Kumar (Ed.), *E-democracy for smart cities* (pp. 197–257). Springer.

Levin, K. A. (2006). Study design III: Cross-sectional studies. *Evidence-Based Dentistry, 7*(1), 24–25.

Melnikovas, A. (2018). Towards an explicit research methodology: Adapting research onion model for futures studies. *Journal of Futures Studies, 23*(2), 29–44.

Pascaris, A., Schelly, C., Burnham, L., & Pearce, J. M. (2021). Integrating solar energy with agriculture: Industry perspectives on the market, community, and socio-political dimensions of agrivoltaics. *Energy Research & Social Science, 75*, 102023.

Sandelowski, M. (2000). Combining qualitative and quantitative sampling, data collection, and analysis techniques in mixed-method studies. *Research in Nursing & Health, 23*(3), 246–255.

Saunders, M., Lewis, P., & Thornhill, A. (2007). *Research methods. Business students 4th edition* (8th ed.). Pearson Education Limited.

Shahsavari, A., & Akbari, M. (2018). Potential of solar energy in developing countries for reducing energy-related emissions. *Renewable and Sustainable Energy Reviews, 90*, 275–291.

Sim, L. C. (2023). Renewable energy and governance resilience in the gulf. *Energies, 16*(7), 3225.

Wani, T. A., & Ali, W. S. (2015). Innovation diffusion theory. *Journal of General Management Research, 3*(2), 101–118.

Chapter 6
From Ambition to Action: Current Progress of Indian Corporates Towards Net Zero Targets

Mahalakshmi Konar, Taufiq Khan, Mihir Herlekar, Jogitha Kanapally, and Gayatri Barabde

Abstract Businesses must reduce their Greenhouse Gas (GHG) emissions due to the growing evidence of climate change, which is manifested in the increased frequency of natural disasters, disruptions in supply chain and scarcity of resources. Therefore, this study explores whether the Indian companies are making efforts to reduce their emissions through Science-Based Targets Initiatives (SBTi). Studies suggests that energy sector, industry, Agriculture, Forestry, and Other Land Use (AFOLU), Transport and Buildings are the major contributors of climate change, and this study aims to know which is the leading sector that is committed to reduce their emissions or has taken initiatives to set target to achieve net zero. Data was retrieved from the SBTi dashboard for the month of December 2023 and publicly available sustainability, or integrated reports (F.Y. 2022–23) were used during this study. Findings suggested that the transport sector and mining sub-sector were the least ones to commit and set emissions reduction targets. Companies of the textile industry were the highest to commit but only few of them set emission reduction targets. On the other hand, a smaller number of companies of professional industry committed but they were leading sectors to set targets for emission reduction. Companies are highly focusing on switching to renewable energy as the energy sector is the major contributor of

M. Konar
Institutional Shareholder Services, ESG Ratings Analyst, Mumbai, India

T. Khan
ESG Executive, CRISIL Limited, Mumbai, India

M. Herlekar (✉) · G. Barabde
Department of Environmental Science, The Institute of Science, Mumbai, India
e-mail: tevs.mihir@iscm.ac.in

G. Barabde
e-mail: gayatri@iscm.ac.in

J. Kanapally
Tikao Technologies, Navi Mumbai, India
e-mail: jogitha@tikao.in

© The Author(s), under exclusive license to Springer Nature Switzerland AG 2026

J. Dsilva (ed.), *Redefining Sustainability*, https://doi.org/10.1007/978-3-031-88376-7_6

GHG emissions caused by coal combustions. Only a few corporations are trying to optimize their emissions generated from their process and value chain.

Keywords Greenhouse gas emissions · Climate change · SBTi · Indian companies

Introduction

According to IPCC reports, if the global temperatures rise by more than 2 °C, there would be irreversible climate change effects. In pursuit of mitigating these consequences, governments and the private sector have actively endeavored to establish and achieve emission reduction goals, with the overarching aim of attaining a state of net zero emissions. This concerted effort is poised to curtail the extent of global warming to 1.5 °C (ISO 2022). The etiology of climate change is multifaceted, and its effects are long-term, uncertain, intricate and challenging to forecast. Notably, the study conducted by Tol (2009) postulates that nations characterized by lower income levels and proximity to the equatorial region are anticipated to endure more pronounced adversities arising from climate change, coupled with a diminished capacity to adapt effectively. Because of their disproportionate emissions, high income nations can be held responsible for the repercussions that low-income countries would face. He also notes in his study that "While it is relatively easy to imagine a disaster scenario for climate change—e.g., involving massive sea level or monsoon failure that could even lead to mass migration and violent conflict—it is not at all easy to argue that climate change will be a huge boost to economic growth".

Corporations are responsible for 40% of global emissions. These corporations initially commit strongly to addressing the climate issue but eventually fade in the "business as usual" scenario (Wright & Nyberg, 2017). This motivated us to study how many Indian companies are committing to reduce their emissions and how many are setting targets to reduce their emissions through SBTi. Firms have devised strategies to alleviate climate change and have also benefited from the reduced energy consumption, but these measures fall short in achieving their absolute emission reductions, and this is referred to as organizational inaction by Slawinski et al. (2016).

Giesekam et al. (2021) scrutinized 81 companies that set targets through SBTi and found that the majority of the companies were on the track to achieving targets, with 21% of them already achieving their targets because they were short-term and the rest of the 35% lagged behind on the set targets. The majority of the companies that participate in SBTi to set targets are found to be located in Europe and North America. Economic incentives prompt companies to set more strenuous targets to reduce emissions and those that adopt external standards like science-based targets exhibit tangible shifts in behavior which becomes a part of collaborative effort to limit global warming (Frieberg et al., 2019).

The number of firms reporting to the Carbon Disclosure Project (CDP) has been found to grow five times from the year 2003 to 2014 (Tang & Demeritt, 2017). SBTi, a joint program of CDP, helps companies to set targets scientifically. These targets operate as a catalyst for carbon mitigation projects, offer recommendations for the selection of suitable mitigation measures, and establish benchmarks for tracking the advancement of mitigation activities (Wang & Sueyoshi, 2018). Findings of the study carried out by Romito et al. (2023) show that firms that participated in SBTi lowered their GHG emissions compared to those which didn't participate.

This study focuses on which industry and sector of India is leading in setting targets to reduce their scope emissions and are committed to reduce their emissions through SBTi. It also compares whether the industries and sectors that are the major contributors of rising GHG emissions are committed toward mitigating climate change.

Methodology

The primary objective of this research is to investigate Indian companies (n = 268) that have formulated specific targets and commitments aimed at reducing their emissions through the implementation of Science-Based Targets initiatives (SBTi). The SBTi dashboard, a publicly accessible platform, facilitates the retrieval of a comprehensive list encompassing companies of varying sizes that have established emission reduction targets or formalized commitments. The dataset extracted from this platform includes pertinent information such as Company name, ISIN, LEI, Near term—target status, Near term—target classification, Near term—target Year, Long term—target status, Long-term target—classification, Long-term target—Year, Net Zero committed, Net Zero Year, Organization type, BA1.5, BA1.5 Date, Location, Region, Sector, Date, Target, Target Classification, and Extension.

Additionally, this research undertakes an examination of the latest sustainability or integrated reports for the fiscal year 2022–23 of publicly listed companies. This facet of the study is necessitated by the unavailability of data pertaining to small and medium enterprises (SMEs) and private companies in the public domain. The objective is to gain insights into the greenhouse gas (GHG) emissions of these companies, discerning the measures they have implemented to curtail their scope emissions. The study period encompasses the month of December 2023, providing a contemporaneous snapshot of the initiatives undertaken by these entities in pursuit of emissions reduction.

Results and Discussions

There were 268 companies in the downloaded excel file that had either committed to cut down or had set target to reduce their emissions. Of these 268 companies, 6 companies were removed from the list by SBTi, 127 companies had committed to reduce their emissions and 135 companies had set targets to reduce their emissions.

The textile industry assumed a notable position in the global emissions landscape, contributing between 6 and 8%, marking it as the second-largest consumer industry (Imran et al., 2023). Notably, the textile, apparels, footwear, and luxury goods sector emerged as the foremost industry in commitment, followed by software and services industry. They were the second leading sector in committing to reducing their emissions. In the case of trading companies and distributors, commercial services and supplies, water transportation, healthcare providers, healthcare technology, forest and paper products where the least number of commitments observed. This deviates from the findings of the IPCC report which states that transport and industry sectors have been contributing to rising GHG emissions across the globe.

While examining companies at the forefront of setting targets for scope emissions reduction distinctive patterns were revealed. Professional service firms emerged as leaders in setting net zero targets, with near term targets well below 1.5 °C. The second leading industry to set net zero targets is software and services, while Construction and Engineering secures the third position. According to the IPCC report, transport is the third largest sector that contributes to global GHG emissions. Surprisingly, we found that only one Indian company in the transportation sector has emissions reduction targets while being the eleventh company by logistics' market capital share listed on National Stock Exchange of India. Companies can adopt strategies such as material efficiency, circularity, electrothermal heating, heat pumps, low GHG emitting fuels to help achieve net zero in mining, manufacturing and light industry (IPCC, 2022). However, only four mining industries have set targets to reduce their emissions.

There are internal and external barriers that lead to ineffective measures toward climate action. External barriers could be regulatory uncertainty, lack of pressure and incentives, complex systems and consumer resistance. Lack of awareness and commitment, lack of knowledge and expertise, lack of resources, short-term planning, cost factor and technology constraints could be the internal barriers affecting the emission reduction measures (Johnson et al., 2023).

Upon reviewing our findings on the study of sustainability and integrated reports of the companies it was observed that all the companies are striving to reduce their scope 2 emissions by switching to renewable energy. Some companies are also trying to monitor their energy consumption and optimize them by using star rated lights and fans while some purchase carbon offsets to become carbon neutral. Pharmaceutical companies are using natural gas and biomass in boilers rather than coal or furnace oil indicating a slight positive change in their operations. Some projects are sequestering carbon through tree plantations. Cement companies are reducing clinker and adopting new technologies to optimize their energy consumption of the processes.

Conclusion

India stands as one of the world's fastest-growing economies, characterized by a burgeoning consumer market, inevitably leading to an upsurge in material and energy consumption. More companies should set targets through SBTi to lower their emissions. Companies cannot achieve net zero by switching to cleaner fuels and energy; rather it should shift focus to reduce its emissions across its value chain, process and material efficiency and change to energy efficient machines. Such measures, undertaken with a long-term perspective, not only augment profitability but also mitigate the environmental footprint of businesses. It is essential to note the scope of this study, which is confined to Indian companies. However, there is potential for broader applicability by extending this research to comprehend global industry endeavors toward achieving net zero emissions. This study can serve as a guideline for the companies belonging to the sectors that are expected to be influenced by Carbon Border Adjustment Mechanism (CBAM). The progress achieved by the corporates can also be correlated or verified with their CDP scores.

References

Freiberg, D., Grewal, J., & Serafeim, G. (2019). Science-based carbon emissions targets. *Social Science Research Network*. https://doi.org/10.2139/ssrn.3804530

Giesekam, J., Norman, J., Garvey, A., & Betts-Davies, S. (2021). Science-based targets: On target? *Sustainability, 13*(4), 1657. https://doi.org/10.3390/su13041657

Imran, S., Mujtaba, M., Zafar, M., Hussain, A., Mehmood, A., Farwa, U. E., Korakianitis, T., Kalam, M., Fayaz, H., & Saleel, C. A. (2023). Assessing the potential of GHG emissions for the textile sector of Pakistan: A baseline study. *Heliyon, 9*(11), e22404. https://doi.org/10.1016/j.heliyon.2023.e22404

Intergovernmental Panel for Climate Change. (2022). Climate change 2022: Mitigation of climate change. In P.R. Shukla, J. Skea, R. Slade, A. Al Khourdajie, R. van Diemen, D. McCollum, M. Pathak, S. Some, P. Vyas, R. Fradera, M. Belkacemi, A. Hasija, G. Lisboa, S. Luz, J. Malley (Eds.) Contribution of Working Group III to the Sixth Assessment Report of the Intergovernmental Panel on Climate Change. Cambridge University Press. https://doi.org/10.1017/9781009157926

Johnson, M., Rötzel, T. S., & Frank, B. (2023). Beyond conventional corporate responses to climate change towards deep decarbonization: A systematic literature review. *Management Review Quarterly, 73*(2), 921–954. https://doi.org/10.1007/s11301-023-00318-8

Romito, S., Vurro, C., & Pogutz, S. (2023). Joining multi-stakeholder initiatives to fight climate change: The environmental impact of corporate participation in the science based targets initiative. *Business Strategy and the Environment*. https://doi.org/10.1002/bse.3639

Slawinski, N., Pinkse, J., Busch, T., & Banerjee, S. B. (2016). The role of short-termism and uncertainty avoidance in organizational inaction on climate change. *Business & Society, 56*(2), 253–282. https://doi.org/10.1177/0007650315576136

Tang, S., & Demeritt, D. (2017). Climate change and mandatory carbon reporting: Impacts on business process and performance. *Business Strategy and the Environment, 27*(4), 437–455. https://doi.org/10.1002/bse.1985

The International Organization for Standardization (ISO). (2022). *Net zero guidelines accelerating the transition to net zero.* https://www.iso.org/netzero#:~:text=The%20Net%20Zero%20Guidelines%20

Tol, R. S. (2009). The economic effects of climate change. *Journal of Economic Perspectives, 23*(2), 29–51. https://doi.org/10.1257/jep.23.2.29

Wang, D., & Sueyoshi, T. (2018). Climate change mitigation targets set by global firms: Overview and implications for renewable energy. *Renewable & Sustainable Energy Reviews, 94*, 386–398. https://doi.org/10.1016/j.rser.2018.06.024

Wright, C., & Nyberg, D. (2017). An inconvenient truth: How organizations translate climate change into business as usual. *Academy of Management Journal, 60*(5), 1633–1661. https://doi.org/10.5465/amj.2015.0718

Chapter 7
The Adoption of Electric Boats in Whale-Watching Tour Operators in Vancouver Island, British Columbia, Canada

Yousif Tulefat

Abstract Employing Rogers' Theory of Adoption as the theoretical framework, this research explores the adoption of electric boats among whale-watching tour operators in Vancouver Island, British Columbia, Canada, Iceland, and Norway. The study aims to understand the factors influencing tour operators' decision-making in embracing electric boats for eco-friendly whale-watching experiences. Rogers' Theory of Adoption serves as a lens to analyze the diffusion of innovation within the maritime tourism industry, mainly focusing on electric boats' attributes and the operators' perceptions toward this sustainable technology. The research incorporates one qualitative method which are interviews, to gather insights from whale-watching tour operators. The comparative analysis between Vancouver Island and Nordic countries provides a nuanced understanding of regional variations in adoption patterns. By shedding light on the dynamics of electric boat adoption, this research contributes to the sustainable transformation of the whale-watching industry. The findings will inform policymakers, boat manufacturers, and tour operators on strategies to accelerate the transition to environmentally friendly practices, fostering a harmonious coexistence between marine tourism and ecological conservation.

Keywords Rogers' theory of adoption · Whale watching · Electric boats · Methodology · Results

Introduction

This abstract explores the adoption of electric boats among whale-watching tour operators. The geographical context of the study is in three locations: Norway, Iceland, and Canada. Whale-watching tour operators in Iceland and Norway have already adopted and transitioned from diesel-fueled to electric boats. Vancouver Island, British

Y. Tulefat (✉)
Nanaimo, Canada
e-mail: yousiftulefat@gmail.com

© The Author(s), under exclusive license to Springer Nature Switzerland AG 2026
J. Dsilva (ed.), *Redefining Sustainability*, https://doi.org/10.1007/978-3-031-88376-7_7

Columbia, Canada's whale-watching tour operators don't have electric boats. The burgeoning concern for environmental sustainability, coupled with the unique challenges faced by marine ecotourism operators, has compelled an examination of alternative technologies to traditional fuel-dependent vessels. This study seeks to elucidate the factors influencing the adoption of electric boats, a nascent yet promising technology, within this industry. Rogers' Theory of Adoption, a well-established model in the field of innovation diffusion, provides a structured lens through which to understand the dynamics of technology adoption (Rogers, 1983). The theory's core elements—innovation, communication channels, time, social system, and perceived attributes—are employed to analyze the adoption patterns, challenges, and success stories within the context of electric boat integration (Rogers, 1983). The following section will explore the research questions and the chosen type of methodology.

Methodology

There are two main research questions that the paper will be examined with different aspects of research with geographical contexts.

Research Question 1: What factors influence the adoption and diffusion of electric boats among whale-watching tour operators in Iceland and Norway, and how does this adoption impact sustainability and innovation in the ecotourism industry?

This research question focuses on adopting electric boats and their impact on sustainability and innovation. It explicitly looks at two countries, Iceland and Norway, known for their whale-watching tours and where some operators have adopted electric boats. Overall, this research question is designed to study why and how whale-watching tour operators in Iceland and Norway are adopting electric boats and the potential impacts of this adoption on sustainability and innovation within the ecotourism industry.

Research Question 2: How does using adoption theory inform the decision-making process of whale-watching tour operators in British Columbia when considering the adoption and transition of electric boats as a sustainable alternative?

This research question centers on adopting electric boats as a sustainable alternative by explicitly focusing on British Columbia whale-watching tour operators. The "adoption theory" is a theoretical framework that will be applied to understand the decision-making process of these tour operators. Incorporating adoption theory from Iceland and Norway adds a comparative dimension to the research. Overall, the research question is well-structured and provides a clear direction for the study. It focuses on the practical application of adoption theory in a real-world context, offering valuable insights into the decision-making processes of whale-watching tour operators in British Columbia.

Case studies are a research methodology that involves in-depth analysis of complex issues within their natural context (Crowe et al., 2011). This well-established approach is widely employed across various disciplines, particularly the social sciences, and entails comprehensive investigation into specific events or phenomena

(Crowe et al., 2011). The case study method effectively addresses explanatory queries on "how," "what," and "why," making it an optimal choice for gathering detailed information (Crowe et al., 2011).

In their study, Crowe et al. (2011) categorized case studies into three distinct types: instrumental, collective, and intrinsic. This paper will focus on the collective case study approach, which involves analyzing multiple examples concurrently or chronologically to present a more comprehensive overview of a particular issue (Crowe et al., 2011).

The thesis will be a conducted collective case study. There are four steps of conducting a collective case study (Yin, 2014). The first step is developing a theory, which is Rogers's thesis on adoption theory (Yin, 2014). The second step is the selection of cases in which the researcher is researching the adoption of electric boats for whale-watching tour operators in British Columbia and Nordic countries. The data collection protocol will be designed using semi-structured interviews with the whale-watching tour operators (Yin, 2014). The third step is the first case study, which is the whale-watching tour operators in the Nordic countries that have already adopted electric boats, and there will be a report on them. The second case study is the British Columbia whale-watching tour operators who want to adopt electric boats, and there will be a report on them (Yin, 2014). Later on, the researcher will do the last step, which is the cross-sectional analysis of the case studies, and discuss it in the discussion part of the paper (Yin, 2014).

The data collection will be conducted through online and face-to-face interviews. The online interviews will be conducted on Zoom, a shared and live video program you can download to your computers or tablets. The live videos can also be recorded with the caller's and callee's permission (Archibald et al., 2019). Zoom-recorded videos can be saved onto your devices. Using Zoom as a research tool for data collection can make it less expensive for the researcher to travel to Nordic and other British Columbian whale-watching tour operators' locations (Archibald et al., 2019). However, one of the interviews is locally based, and the researcher will conduct an interview face-to-face with the owner of the whale-watching tour operator. Both face-to-face and Zoom interviews will be semi-structured interviews. According to Ryan et al. (2009), this is more reasonable for the qualitative research method because the open-ended questions can have an open discussion between the interviewer and interviewee. The results of the data analysis are more prosperous and more rigorous.

Expected Findings

The interviews will be conducted in February 2024; however, we can intend assumptions from both geographical contexts. For the Nordic countries, whale-watching tour operators have already adopted electric boats in their operations. The researcher will ask questions about "How they adopted the electric boats?" "what were the impacts of electric boats?" and "Why were the motivations to transition to electric boats?". Making it an optimal choice for gathering detailed information (Crowe et al., 2011) since this is a case study method. For the British Columbian whale-watching tour operators, the researcher will be asking the same theme questions but with "How would you adopt the electric boats," "What strategies would to find solutions to barriers to adopt the electric boats," and "Why would you adopt the electric boats" they want to adopt the electric boats. The data analysis plan for the interview results will be thematic analysis. Thematic is "a method for analyzing qualitative data that entails searching across a data set to identify, analyze, and report repeated patterns. It is a method for describing data, but it also involves interpretation in the processes of selecting codes and constructing themes" (Kiger & Varpio, 2020). In the results, there will be "patterns" of activities of whale-watching tour operators or "data" of numbers and statistics or "frequency" of keywords that are important to the research.

Conclusion

In conclusion, this study will utilize Rogers' Theory of Adoption and offer valuable insights into the dynamics of electric boat adoption within whale-watching tour operators in Vancouver Island, British Columbia, Canada, and Nordic countries. The findings will underscore the multifaceted nature of the decision-making process, influenced by factors such as perceived advantages, economic considerations, and regulatory landscapes. By understanding these dynamics, whale-watching tour operators can formulate targeted strategies to accelerate the transition to sustainable practices in the marine tourism sector. This research contributes to the broader discourse on the adoption of eco-friendly technologies, providing a foundation for informed policymaking and industry collaboration to foster a more environmentally responsible future for whale-watching operations.

References

Archibald, M. M., Ambagtsheer, R. C., Casey, M. G., & Lawless, M. (2019). Using Zoom videoconferencing for qualitative data collection: Perceptions and experiences of researchers and participants. *International Journal of Qualitative Methods, 18*, 1–8. https://doi.org/10.1177/1609406919874596

Crowe, S., Cresswell, K., Robertson, A., Huby, G., Avery, A., & Sheikh, A. (2011). The case study approach. *BMC Medical Research Methodology, 11*(100), 1–9. https://bmcmedresmethodol.biomedcentral.com/https://doi.org/10.1186/1471-2288-11-100

Kiger, M. E., & Varpio, L. (2020). Thematic analysis of qualitative data: AMEE guide no. 131. *Medical Teacher, 42*(8), 846–854. https://doi.org/10.1080/0142159X.2020.1755030

Rogers, E. M. (1983). In F. F. Shoemaker (Ed.), *Diffusion of innovations*. The Free Press. https://teddykw2.files.wordpress.com/2012/07/everett-m-rogers-diffusion-of-innovations.pdf

Ryan, F., Coughlan, M., & Cronin, P. (2009). Interviewing in qualitative: research: The one-to-one interview. *International Journal of Therapy and Rehabilitation, 16*(6), 309–314. https://doi.org/10.12968/ijtr.2009.16.6.42433

Yin, R. K. (2014). *Case study research design and methods* (5th ed.). SAGE Publications Inc. https://handoutset.com/wp-content/uploads/2022/06/Case-Study-Research-Design-and-method-Robert-K.Yin_pdf

Chapter 8
The Role of Urban Green Infrastructure on Air Quality: The Case of Dubai

Makbulenur Onur and Yasemin Nielsen

Abstract The increasing urbanization, changing population distribution, rapid depletion of natural resources, and declining air quality in cities pose numerous challenges today. Green Infrastructure (GI) practices are developed to address these issues by making cities sustainable, improving air quality, and creating healthy urban environments adaptable to climate change. However, accurately implementing GI practices is complex. This study aims to provide design recommendations by analyzing which GI application improves urban air quality on a micro scale and how they interrelate, specifically in hot climate conditions. Understanding which GI features enhance air quality in different seasons and provide year-round sustainable solutions is crucial. The study's results, based on measurable standards and design scenarios, will guide climate-sensitive and purposeful design decisions. Fieldwork will involve air quality measurements using ArcMap grid networks at specified coordinates. Comparing harmful emissions (CO, CO_2, PM10, PM2.5) across regions and statistically demonstrating the positive correlation of GI systems with certain seasons will be emphasized. These results are expected to serve as exemplary for future projects.

Keywords Micro-air quality · Green infrastructure · Air quality index

Introduction

The world is experiencing a significant consumption rate brought about by the changing supply and demand balance. (Rienow et al., 2022; Olabi & Abdelkareem, 2022). Due to humanity's incessant emission production for its own well-being, the atmosphere is becoming polluted, posing a threat to the future. This consumption rate disrupts numerous balances in the universe. Air pollution and rising temperatures are

M. Onur (✉)
Karadeniz Technical University, Trabzon, Türkiye
e-mail: mnurbekar@ktu.edu.tr

Y. Nielsen
Heriot Watt University, Dubai, United Arab Emirates
e-mail: yasemin.Nielsen@hw.ac.uk

© The Author(s), under exclusive license to Springer Nature Switzerland AG 2026

J. Dsilva (ed.), *Redefining Sustainability*, https://doi.org/10.1007/978-3-031-88376-7_8

just a few of these issues. With the increase in air pollution in global climate change, monitoring “air quality change” has become crucial. This monitoring interprets the impact of emissions on human health based on pollutant parameter concentrations in the atmosphere through various measurement methods (Ayres et al., 2006; Ozsoy, 2023).

Green infrastructure practices reduce carbon dioxide levels in the atmosphere and filter harmful particles, providing clean air. This can contribute to the reduction of health problems such as respiratory diseases. However, the fundamental question here is “which green infrastructure systems” are oriented toward improving micro-air quality. Many green infrastructure systems are currently being implemented but determining which one should be used and where each application is suitable remains a complex issue. Despite being known as one of the best ways to control climate conditions, improve air quality, and reduce urban heat effects, green infrastructure applications must be appropriate and correctly implemented to be a solution to the problem (Saaroni et al., 2018; Marando et al., 2022; Iungman et al., 2023).

While numerous scientific studies argue that green infrastructure, sustainable applications could be a solution to improve air quality (Pugh et al., 2012; Berardi et al., 2013; Tallis et al., 2015; Hewitt et al., 2020; Badach et al., 2022; Evans et al., 2022; Wu et al., 2022; Corada Perez, 2023; Borisova et al., 2023; Jin et al., 2024), there is a lack of literature pinpointing which features of green infrastructure applications play a better role in improving air quality (Tomson et al., 2021; Wu et al., 2022). Therefore, this study will examine the relationship between pollutants and green infrastructure systems.

The study area selected for this research includes the JLT Almas Tower and its surroundings as the measurement area and the JLT Park as the working area. Measurement stations will be determined in the study area using ARCmap software, and measurements of pollutants such as CO, CO_2, TVOC, HVOC, PM2.5, and PM10 will be consistently taken from the same locations. The study aims to identify the connection between pollutant ratios in measurement networks and green infrastructure applications. Thus, it will be determined which green infrastructure application positively affects air quality when implemented.

Hypothesis of the Study

The hypothesis that “there is a strong correlation between green infrastructure initiatives and air quality, and that proper green infrastructure projects can improve micro-air quality” forms the basis of this study. The sub-hypotheses of the study are as follows,

- There is a relationship between the type of plant used and air quality.
- Green infrastructure practices affect air quality.

Fig. 8.1 Study area (Google ProMap Data)

Methodology

This study covers a 12-month period. In the first stage of the study, aerial photographs of both study areas (Fig. 8.1) will be defined in the ArcGIS projection system, and maps will be created. Using Geographic Information System (GIS) and ArcMap software, the coordinates of the scaled aerial photograph will be determined to enable emission measurements, and a chart will be generated. This process will assign specific "coordinates" to each frame, allowing for the verification of the correct location using GPS during fieldwork. The findings will be recorded in detection charts and compared with the measurements at the end of the study. Measurements will be taken three times on the designated day at different hours (09:00, 15:00, 21:00). When determining the measurement frequencies, the World Health Organization's Global Air Quality Guidelines published in 2021 will be consulted (WHO, 2021). "Air quality index" maps will be created based on the measurements. CBS (Geographic Information System), ArcGIS, and Photoshop software will be used to generate these maps. Data within the scope of the study will be analyzed using the SPSS program. Tests will be applied to the analysis data to reveal relationships between them. All these data will be supported with statistical information.

Within the scope of the study, measurements were determined using Air Quality Meter. Although measurements for other months of the study continue, the measurements between "18 December and 18 December" have been completed.

Findings

The findings obtained from all conducted studies will be as follows:

- Measurement data from the selected stations within both study areas,
- Measurement findings for each station regarding CO, CO_2, TVOC, HVOC, PM2.5, PM10,
- Green infrastructure applications for each station,

- Vegetation findings for each station,
- Findings related to the types of green infrastructure applications present at each measurement station,
- Micro-air quality index map findings.

All these findings will be identified within two different study areas. The project, spanning a 12-month period, commenced on December 18, 2023, and is still ongoing. This study will discuss the findings and objectives of the project at its current stage. Each finding serves as a substrate data for the subsequent phase.

Conclusion

The results of the study are expected to yield design scenarios created with measurable standards, serving as a guiding resource proposing climate-sensitive and, most importantly, purpose-oriented design decisions. Microclimate index maps and the relationship between green infrastructure applications generated in the study will provide foundational data for future research. This is crucial as the answer to the question of which application currently improves air quality better remains unclear. It is envisioned that this study will serve as an example in addressing this question.

Given that the study's data relies on a statistical and measurement-based process, the outcomes will be supported by numerical data. Maps for each pollutant will be generated, and statistical comparisons will be made at the end of the study. The study is currently in its second month, and within the scope of this work, the goals and findings obtained during this period will be discussed.

The study is ongoing for the remaining months. The data from the 1st month is presented in this study. The relationship between green infrastructure works and micro-air quality is supported even with the data from the first month. Micro-air quality in the first study area varies between 65 and 199 $\mu g/m^3$. In the second study area, it varies between 64 and 76 $\mu g/m^3$. There is no water effect in the first study area, while there is in the second. Despite this, the air quality in the first study area is lower. The high presence of broad-leaved plants in the second study area has significantly increased the air quality. One of the important findings is that the presence of broad-leaved trees is high in places where measurements are good. Another important finding is that air quality is low in places with artificial ponds.

Acknowledgements This study is supported within the framework of the project with ID number 1059B192300479, funded by the 2219-International Postdoctoral Research Fellowship Program for Turkish Citizens (Duration: 18.12.2023–18.12.2024). Partner Universities in the Project: Karadeniz Technical University & Heriot Watt University–Dubai.

References

Ayres, J., Maynard, R., & Richards, R. (2006). *Air pollution and health* (pp. 9–14). Imperial College Press.

Badach, J., Szczepański, J., Bonenberg, W., Gębicki, J., & Nyka, L. (2022). Developing the urban blue-green infrastructure as a tool for urban air quality management. *Sustainability, 14*(15), 9688.

Berardi, U., GhaffarianHoseini, A. H., & Hoseini, G. (2013). State-of-the-art analysis of the environmental benefits of green roofs. *Journal of Applied Energy, 115*, 411–428.

Borisova, B., Dimitrov, S., Ihtimanski, I., Semerdzhieva, L., Valchev, S., & Iliev, M. (2023, September). Geospatial localization analysis of green infrastructure development assumptions in urbanized areas concerned by air quality: the case of Sofia Municipality, Bulgaria. In *Ninth International Conference on Remote Sensing and Geoinformation of the Environment (RSCy2023)* (Vol. 12786, pp. 282–298). SPIE.

Corada Perez, K. A. (2023). *Towards a holistic understanding of the role of green infrastructure in improving urban air quality*. PhD. Thesis Department of Centre for Environmental Policy.

Evans, D. L., Falagán, N., Hardman, C. A., Kourmpetli, S., Liu, L., Mead, B. R., & Davies, J. A. C. (2022). Ecosystem service delivery by urban agriculture and green infrastructure—a systematic review. *Ecosystem Services, 54*, 101405.

Hewitt, C. N., Ashworth, K., & MacKenzie, A. R. (2020). Using green infrastructure to improve urban air quality (GI4AQ). *Ambio, 49*, 62–73.

Jin, M. Y., Zhang, L. Y., Peng, Z. R., He, H. D., Kumar, P., & Gallagher, J. (2024). The impact of dynamic traffic and wind conditions on green infrastructure performance to improve local air quality. *Science of the Total Environment, 917*, 170211.

Iungman, T., Cirach, M., Marando, F., Barboza, E. P., Khomenko, S., Masselot, P., … Nieuwenhuijsen, M. (2023). Cooling cities through urban green infrastructure: A health impact assessment of European cities. *The Lancet, 401*(10376), 577–589.

Marando, F., Heris, M. P., Zulian, G., Udías, A., Mentaschi, L., Chrysoulakis, N., … Maes, J. (2022). Urban heat island mitigation by green infrastructure in European functional urban areas. *Sustainable Cities and Society, 77*, 103564.

Olabi, A. G., & Abdelkareem, M. A. (2022). Renewable energy and climate change. *Renewable and Sustainable Energy Reviews, 158*, 112111.

Ozsoy. (2023). *Development of different methods in the assessment of air quality*. Karabuk University Institute of Graduate Studies, Department of Mechanical Engineering, PhD Thesis.

Pugh, T. A., MacKenzie, A. R., Whyatt, J. D., & Hewitt, C. N. (2012). Effectiveness of green infrastructure for improvement of air quality in urban street canyons. *Environmental Science & Technology, 46*(14), 7692–7699.

Rienow, A., Kantakumar, L. N., Ghazaryan, G., Dröge-Rothaar, A., Sticksel, S., Trampnau, B., & Thonfeld, F. (2022). Modelling the spatial impact of regional planning and climate change prevention strategies on land consumption in the Rhine-Ruhr Metropolitan Area 2017–2030. *Landscape and Urban Planning, 217*, 104284.

Tallis, M.J., Amorim, J.H., Calfapietra, C., Freer-Smith, P., Kotthaus, S., Toscano, P.,... Grimmond, S. (2015). The impacts of green infrastructure on air quality and temperature. In *Handbook on Green Infrastructure* (pp. 30–49). Edward Elgar Publishing.

Tomson, M., Kumar, P., Barwise, Y., Perez, P., Forehead, H., French, K., … Watts, J. F. (2021). Green infrastructure for air quality improvement in street canyons. *Environment International, 146*, 106288.

World Health Organization. (2021). *WHO global air quality guidelines*. Arrival at: May 7, 2023. https://apps.who.int/iris/bitstream/handle/10665/345329/9789240034228-eng.pdf

Wu, H. W., Kumar, P., & Cao, S. J. (2022). Implementation of green infrastructure for improving the building environment of elderly care centres. *Journal of Building Engineering, 54*, 104682.

Saaroni, H., Amorim, J. H., Hiemstra, J. A., & Pearlmutter, D. (2018). Urban Green Infrastructure as a tool for urban heat mitigation: Survey of research methodologies and findings across different climatic regions. *Urban Climate, 24*, 94–110.

Chapter 9
Investigating the Barriers in Photovoltaic (Solar) Energy Generation and Utilization in Dubai, United Arab Emirates

Keyur Athaide and Mohamed Irfan Shaikh

Abstract Global warming demands urgent action, especially in utilizing clean energy. This study identifies barriers to solar energy use in Dubai, highlights its benefits, addresses knowledge gaps, and offers solutions. Using Saunders' research onion methodology, data was collected through surveys. Despite a knowledge gap, findings show 80% + respondents are engaged with solar technology. Key barriers and solutions were identified. Statistical analysis revealed a strong positive correlation between key variables. The study's objectives were met, and implementing the recommendations can lead to significant consumer savings, industry growth, and environmental benefits, promoting a cleaner and greener future.

Keywords Photovoltaic energy generation · Clean energy · Solar technology · Environmental benefits

Introduction

Dubai has an average of 335.4 sunshine days a year (Weather Atlas, 2022) and it would be expected that the most extensively used form of renewable energy in Dubai is solar or photovoltaic (PV) grids. However, the proposed solar power generation of merely 5 gWh by 2030 (TDRA, 2022b), translates to a contribution of only 0.011% based on the latest annual energy consumption of 45,712 gWh (Dubai Electricity & Water Authority, 2021). The aim of this research study is to identify and investigate the various barriers that have been restricting the generation and utilization of renewable energy and to determine its impact on consumers in perspective of

K. Athaide
Dry Docks World, Dubai, United Arab Emirates

M. I. Shaikh (✉)
SEE Institute, The Sustainable City, Dubai, United Arab Emirates
e-mail: irfan@seeinstitute.ae

© The Author(s), under exclusive license to Springer Nature Switzerland AG 2026

J. Dsilva (ed.), *Redefining Sustainability*, https://doi.org/10.1007/978-3-031-88376-7_9

the United Nations Climate Agreement. This study also aims to highlight the benefits and importance of utilizing this energy source while considering the following challenges.

- Misconceptions due to limited knowledge available about green/clean/renewable energy.
- The consumer mindset of choosing convenience or social responsibility with reference to the United Nations Sustainable Development Goals & Climate Action.
- Cost effective models in the utilization of photovoltaic grid installations.

This research will help address the knowledge gap in the utilization of solar energy (photovoltaic grids) in Dubai and create a sense of responsibility toward aligning our thought process with the United Nations Sustainable Development Goals for Climate Action. The outcome of this study will also provide an opportunity to rethink our strategy and minimize the harmful after-effects that could befall us, should we ignore our responsibility and commitment to our future generations.

Research Objectives

1. To identify the barriers and knowledge gaps among consumers and their choice of convenience versus social responsibility.
2. To review and understand the present level and extent of utilization of photovoltaic grids in Dubai.
3. To create a feasible approach/plan to overcome barriers in the utilization of solar energy (PV grids), thus increasing the use of renewable energy and leading to conservation of fossil fuels in a limited resource industry.

Research Questions

1. What is the extent of photovoltaic/solar energy utilization in Dubai?
2. What are the barriers restricting us from accepting and proceeding with solar/photovoltaic installations as a responsibility toward our prolonged existence on the planet ?
3. What methodology can an underdeveloped industry utilize to overcome barriers and install photovoltaic grids to optimize the use of clean, green, and renewable energy resources ?

Literature Review and Gap Analysis

See Table 9.1.

Table 9.1 GAP Analysis—Knowledge gap summary addressed in this study

Journal Name	Authors / Researchers	Title / Focus of Research	Research Aim	Study Method	Findings	Comments / Knowledge Gap
Materials Today	C.M. Vivek / P. Ramkumar / P.K. Srividhya / M. Sivasubramanian	Recent strategies and trends in implanting of renewable energy sources for sustainability	To assess renewable energy sources in various countries by highlighting their performance towards implementation of new strategies and their performance towards energy sustainability	Mixed Method	Development of solar & wind energy sources is promising but limited by cost & technological requirements in developing countries.	Global renewable energy utilization trends and implementation strategy
Journal of Cleaner Production	Iqra Javid / Aditya Chauhan / Sahil Thappa / S.K. Verma / Y. Anand / A. Sawhney / V.V. Tyagi / S. anand	Futuristic decentralized clean energy networks in view of inclusive-economic growth & sustainable society	To review decentralized-distributed energy networks and the need for energy planning, design configurations & implementation models.	Mixed Method	Radical shift towards applied renewable resources, integration & applications. Major challenges and shortcomings are due to lack of information and planning.	Status of renewable energy development, its importance, challenges, effect on environment & feasibility
Journal of Cleaner Production	Azmat Gani	Fossil fuel energy and environmental performance in an extended STIRPAT model	To examine the supply - side effect of energy generation from fossil fuels on environmental quality.	Quantitative Method	There is a negative correlation of energy from fossil fuel combustion due to which countries need to invest in policies that better protect the environment	The increasing demand for fossil fuel energy over the last few decades & its effect on global warming.
Sustainability	Vincenzo Basile / Francesca Loia / Roberto Vona	Towards a Sustainable Decommissioning of Offshore Platforms in the Oil & Gas Industry	To address issues about decommissioning platforms and the importance to combine economic & environmental sustainability	Qualitative Method	Diverse concepts can be considered with readaptation of these assets along with a life-cycle cost benefit.	Government legislations, social responsibility & climate change regulations and complications with harvesting & storing green energy.
Journal of Building Engineering	Joohyun Lee / Mardelle Shepley	Benefits of solar PV systems for low-income families in social housing of Korea: Renewable energy applications as solutions to energy poverty	To investigate the effectiveness of Korean government's sustainability energy policy providing PV systems for low-income households.	Mixed Method	Increased user awareness is important for the successful implementation of this project along with considering structural modifications to improve efficiency.	Importance of customer knowledge for successful implementation of renewable projects.
Sustainability	Izabela Rampasso / Geraldo Filho / Rosley Anholon / Robson de Araujo / Gilson Lima / Luis Zotes / Walter Filho	Challenges presented in the implementation of sustainable energy management	To identify the challenges found by organisations during implementation of ISO 50001:2011 with regard to sustainable development.	Qualitative Method	The ISO 50001 guideline lacks implementation strategy but is important to understand the implications & consequences. Industrial policies are required to ensure implementation.	Understanding the new methods of generating clean energy and the advantages & limitations of each method.

(continued)

Table 9.1 (continued)

European Journal of Sustainable Development Research	Arslan Habib / Chan Sou	Analytical Review on the Trends and Present Situation of Large-scale Sustainable Energy Storage Technology	To identify various storage methods being implemented and considered for renewable and new energy technologies and the challenges faced.	Qualitative Method	New technologies and development trends for energy storage, present storage challenges of the energy industry & the opportunities it presents.	Various methods of energy storage, transportation and resource locations.
Renewable and Sustainable Energy Reviews	Vasco Brummer	A comparative literature review of Community Energy in the UK, Germany & the USA, the benefits it provides for society & the barriers it faces.	To sum up knowledge in the field of Community Energy with focus on benefits and the barriers energy projects face in their respective countries.	Qualitative Method	The policy framework plays a major role in Community Energy (CE) movement. Policy makers should keep in mind new regulations, incentives & consequences in the energy sector.	The four key types of barriers & limitations that hinder the development & growth of the renewable energy industry.
Sustainability	Yuehong Lu / Zafar Khan / Manuel Alvarado / Yang Zhang / Zhijia Huang / Md. Imran	A Critical Review of Sustainable Energy Policies for the Promotion of Renewable Energy Sources	To review the sustainable energy policy on promotion of renewable energy & development history in USA, Germany, UK, Denmark & China	Mixed Method	Government & stakeholders need to actively increase & promote policy incentives & control to increase renewable energy adoption.	Controlling & mitigating challenges to increase access to affordable, reliable & adequate energy supply & its impact across various levels.
Renewable Energy Driven Future	D. Sakthivadivel / K. Balaji / D. Dsilva / S. Iniyan / L. Suganthi	Solar Energies Technologies - Principles and applications	To identify various solar energy technological advancements & applications.	Mixed Method	There are numerous methods to generate, collect & utilize solar energy, both directly & indirectly. Each system has varying efficiencies based on the conversion methods and the type of material used in the solar cells.	History of solar energy, its utilization, technological advancements & material used in solar grids.
International Journal of Energy Research	Md. Hayat / Danish Ali / Keitumetse Monyake / Lana Alagha / Niaz Ahmed	Solar energy-A look into power generation, challenges, and a solar-powered future	To review the direct / indirect ways to produce electricity from solar energy and advanced procedures for photovoltaic characterization and performance rating.	Mixed Method	The economics & efficiency of solar cell depends on its material, the relation between spectral responsivity, its power output and current generation have been derived.	Sources of energy, its utilization & solar energy efficiencies based on various types of transformation.
Royal Society of Chemistry	Jinlong Gong / Can Li / Michael Wasielewski	Advances in solar energy conversion	To review & highlight the urgent mission to develop sustainable energy resources due to increasing energy demand & limited global fossil fuels	Qualitative Method	The utilization of solar energy will be a reality & provide a sustainable lifestyle in the future.	Various advanced methods for solar energy conversion and new age technology
ES Energy & Environment	Chen Chen / Xixi Xie / Ming Yang / Ilwoo Seok / Zhanhu Guo / Quinglong Jiang / Grant Wangila / Hang Zhang / Qibin Liu	Recent Advances in Solar Energy Full Spectrum Conversion & Utilization	To review the full spectrum conversion of solar energy and solar thermal applications.	Mixed Method	PV technology is a direct & effective way to utilize solar energy. However, the mismatch between absorption band of solar cells and sunlight restricts its efficiency. Modifications are necessary to improve its efficiency.	How solar energy is converted to electricity & the components involved in its conversion.

(continued)

Table 9.1 (continued)

Energy	Lujean Ahmad / Navid Khordehgah / Jurgita Malinauskaite / Hussam Jouhara	Recent advances & applications of solar photovoltaics & thermal technologies	To access the extent of exploitation of renewable energy sources & incentives to achieve energy efficiency supported by the European regulatory framework & policies.	Qualitative Method	Clear direction from EU & more opportunities to produce clean energy. Innovative technologies are required. Advantages & disadvantages should be considered in the strategy to meet clean energy targets.	Importance of solar energy, its effect on the environment, global urgency in adaptation, industrial growth, increased conversion efficiency and cost efficiency.
International Journal of Hydrogen Energy	Chaouki Ghenai / Tareq Salameh / Adel Merabet	Techno-economic analysis of off grid solar PV/Fuel cell energy system for residential community in desert region	To optimize design & dispatch control strategies of a standalone hybrid renewable power system to meet the desired electric load of a residential community in a desert region.	Mixed Method	The solar PV / Fuel cell off-grid power system integrated with solar based electrolyzer supports renewable energy requirements at a low cost with less energy utilization & zero emissions	Solar energy utilization & it's suitability with regard to the various energy demands for different establishments.
Materials Science for Energy Technologies	K.N. Nwaigwe / P. Mutabilwa / E. Dintwa	An overview of solar power (PV systems) integration into electricity grids	To assist implementation of solar grid integration & provide data for researchers & scientists on the viability of this system.	Qualitative Method	Integration of PV system into grids can reduce transmission & distribution losses and lower generation costs. However, system stability needs to be considered prior to installation.	The functioning of on-grid PV systems, critical components & system arrangement.
Energy Research & Social Science	M. Hazrati / R.J. Heffron	Conceptualising restorative justice in the energy Transition: changing the perspectives of fossil fuels.	To capture the restorative justice and provide a comparative law and case study perspective across Iran, Malaysia & Canada with focus on the oil & gas sector	Qualitative Method	Restorative justice is still a novel concept for the energy justice framework an there is extensive work required to include this within the criminal justice system. The injustices in the oil & gas system can drastically reduce if implemented.	The environmental impact from fossil fuels & from the degradation of solar PV panels.
Renewable Energy	Nicholas Mukisa / Ramon Zamora / Tek Lie	Store-on grid scheme model for grid-tied solar photovoltaic systems for industrial sector applications.	To create a benefit analysis of store-on grid scheme business model for rooftop PV technology in developing countries.	Mixed Method	The study indicates that the proposed model is feasible and should be proposed to other developing countries transitioning towards renewable energy.	Benefits of utilizing solar PV systems for consumers & government.

Source Created by researcher

Table 9.2 Research onion methodology and deduction stages

Philosophy	:	Positivism / Critical Realism / **Interpretivism** / Post-Modernism / Pragmatism
Approach	:	**Deduction** / Abduction / Induction
Methodological Choice	:	**Mono Method Quantitative** / Mono Method Qualitative / Multi-Method Quantitative / Multi-Method Qualitative / Mixed Method Simple / Mixed Method Complex
Strategy	:	Experiment / **Survey** / Archival Research / Case Study / Ethnography / Action Research / Grounded Theory / Narrative Inquiry
Time Horizon	:	Longitudinal / **Cross-Sectional**
Techniques & Procedures	:	Data Collection & Analysis (**Demographic, Descriptive, Statistical T-Test / Linear Regression**)

Source Adapted from Saunders et al. (2016). Created by researcher

Research Methodology

Introduction

Research methodology is a necessary and critical aspect of a dissertation process. It is used to verify the proposed hypothesis within a structured framework and justifies the researchers' philosophy and theories that have been considered to create the foundation of this study. There have been numerous methods proposed by various scholars, with each of them justifying why their method is preferred over the other. The techniques utilized involve theories, diagrams, paradigms, and models that are analyzed in detail by the researcher and quantitatively or qualitatively processed to support the study. This study involves a quantitative research method which unfolds as the basis of this research is presented (Table 9.2).

Method Outline—Research Onion

One of the most preferred and influential methods that has been defined as a systematic process was conducted by Saunders, Lewis, and Thornhill and is popularly known as 'The Research Onion'. Like the layers of the onion, this method helps identify six stages in the research methodology process which progressively takes us toward the

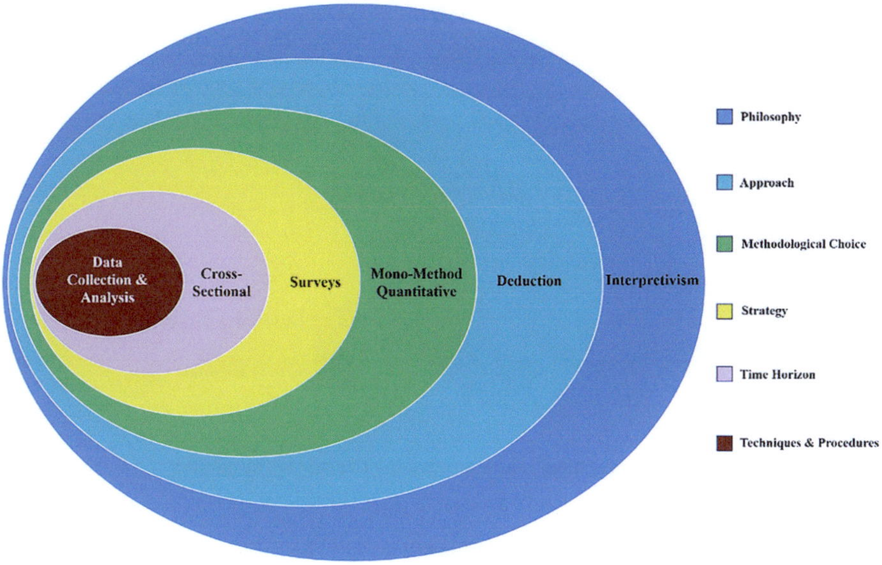

Fig. 9.1 Research plan based on Saunders' research onion. *Source* Created by researcher

core of the onion which is our ultimate objective of the study. The six stages of this methodology that help deduce the path of research from the outermost layer toward the core are represented by philosophy, approach, methodological choice, strategy, time horizon, and techniques and procedures (Fig. 9.1).

Data Collection, Analysis, and Findings

Surveys collect various data types, offering insights into research objectives and testing hypotheses. Statistical analysis uncovers patterns, trends, and approximations for unmeasurable factors, providing clarity on social science topics.

Data was gathered from a random survey in Dubai, targeting respondents aged 21+. Out of 252 responses, 220 were valid based on geographic relevance. The survey included demographic questions, Likert scales, ratings, and multiple-choice questions. The analysis focused on valid responses using descriptive and inferential methods, represented in tables, bar graphs, histograms, and pie charts.

Data Analysis

The analysis is aligned with the study's three objectives:

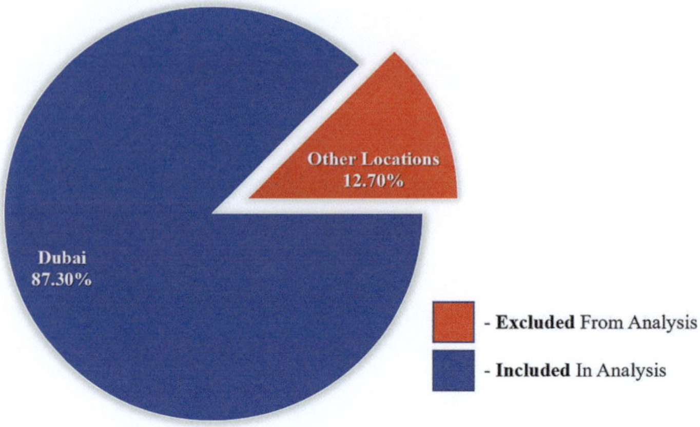

Fig. 9.2 Demography—geographical relevance. *Source* Created by researcher

- Identifying barriers and knowledge gaps.
- Understanding the current use of photovoltaic (PV) grids in Dubai.
- Developing a feasible plan to overcome barriers and increase solar energy use.

Demographic questions (1–5) assessed respondent suitability. Questions 6–8 examined barriers and knowledge gaps. Questions 9–10 evaluated the extent of PV energy use. Questions 11–12 explored feasibility plans to boost solar energy awareness and adoption.

Demographic Analysis

The survey had 252 respondents of which 220 have valid responses (Figs. 9.2, 9.3, 9.4, 9.5, and 9.6).

Descriptive Analysis

The 'Descriptive' analysis provides insight on the various responses received with relevance to each question within this research study. It is the visual representation of data collected. The questions asked and the responses received have been categorized as per the three objectives of this study (Fig. 9.7 and Table 9.3).

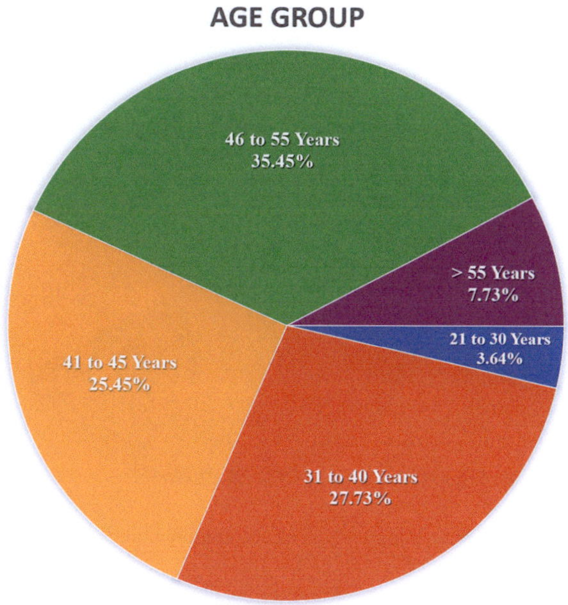

Fig. 9.3 Demography—age. *Source* Created by researcher

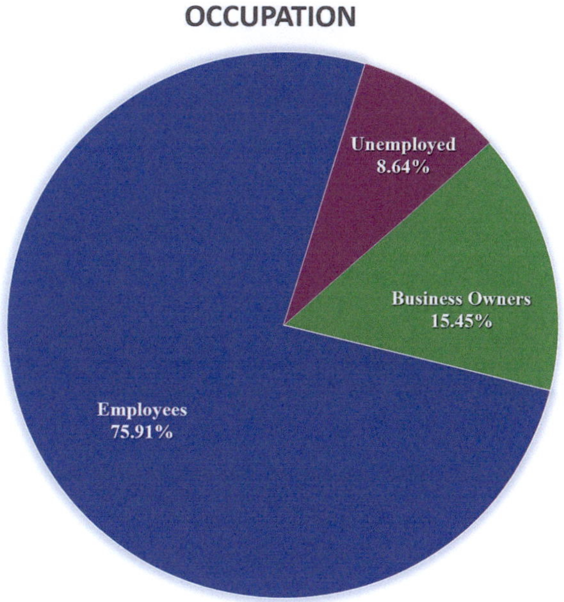

Fig. 9.4 Demography—occupation. *Source* Created by researcher

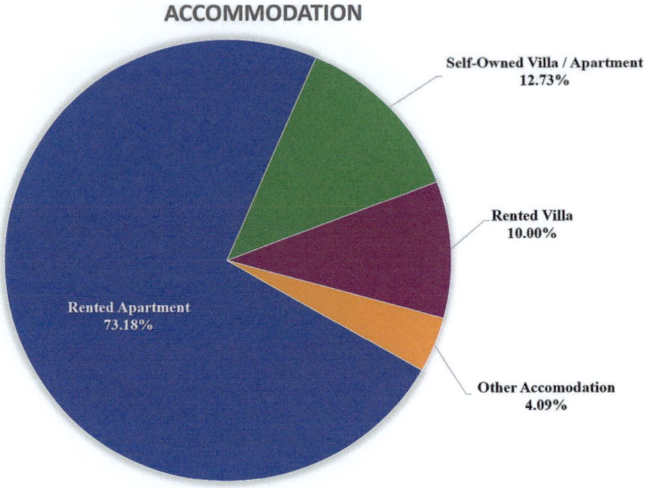

Fig. 9.5 Demography—type of accommodation. *Source* Created by researcher

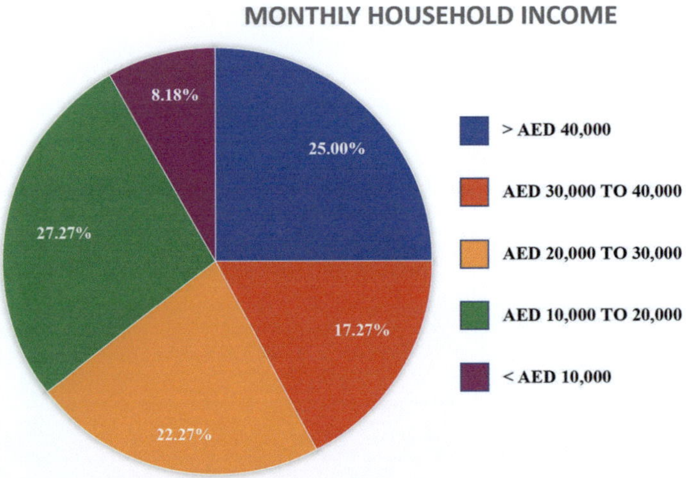

Fig. 9.6 Demography—monthly household income. *Source* Created by researcher

Objective 1—Identifying Barriers/Knowledge Gaps

Question 6:*Renewable, Clean and Sustainable Energy Sources Available in Dubai.*

Question 7:*Most Abundant Energy Source Available in Dubai*

Question 8:*Knowledge and Interest—Renewable, Clean, Green, and Sustainable Energy* (Fig. 9.8).

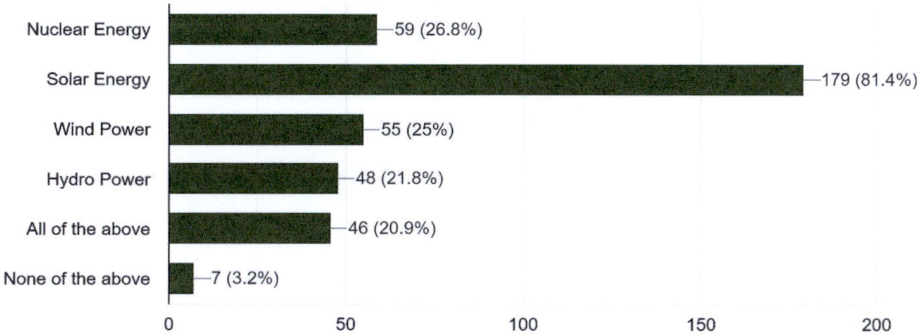

Fig. 9.7 Knowledge gap—energy sources in Dubai. *Source* Created by researcher

Table 9.3 Knowledge gap—most abundant energy sources in Dubai

ENERGY SOURCE	NUMBER OF RESPONDENTS	PERCENTAGE
Wind Power	4	1.82%
Hydro Power	11	5.00%
Solar Energy	133	60.45%
Bio Fuel	14	6.36%
Fossil Fuel	49	22.27%
Nuclear Power	4	1.82%
None of the above	5	2.27%

Source Created by researcher

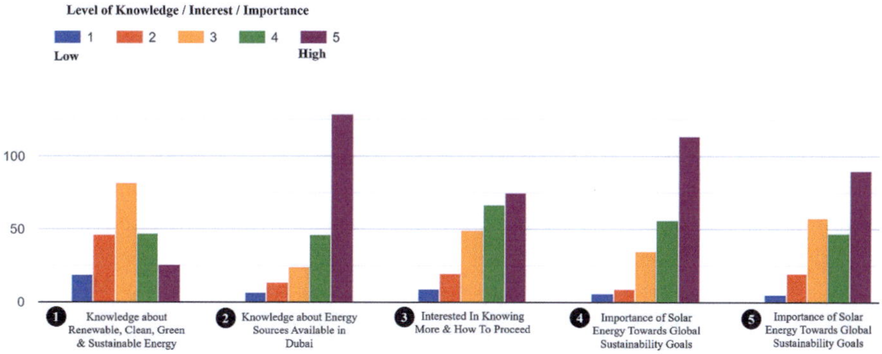

Fig. 9.8 Knowledge gap, interest, and importance. *Source* Created by researcher

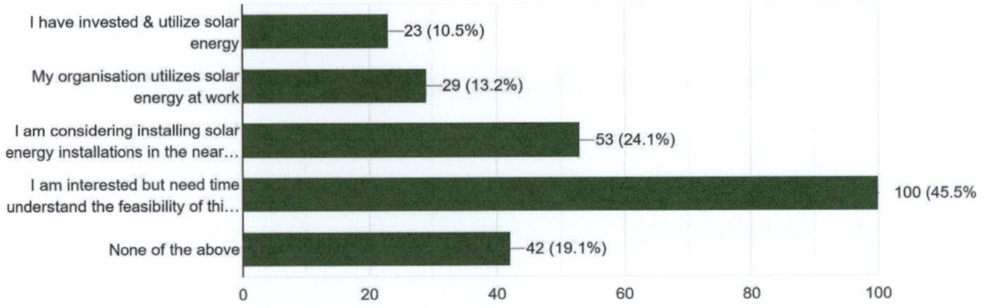

Fig. 9.9 Association and involvement with solar energy. *Source* Created by researcher

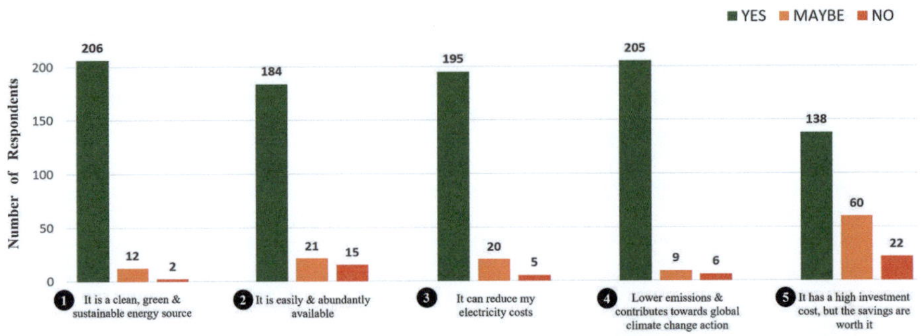

Fig. 9.10 Benefits of utilizing solar energy. *Source* Created by researcher

Objective 2—Extent of Utilization of Solar (PV) Energy

Question 9:*Association and Involvement with Solar Energy* (Fig. 9.9).

Question 10:*Understanding the Benefits of Utilizing Solar Energy* (Fig. 9.10).

Objective 3—Overcoming Barriers and Increasing Solar Energy Utilization

Question 11:*Scenarios that Could Increase Investment and Utilization of Solar Energy*

The above graph is a representation of respondents who were presented with seven different scenarios based on a Likert scale of responses. The respondents were asked if they would consider investing and using solar energy in each of the presented situations. Figure 9.11 indicates that the majority of the respondents were very likely to do so, in each of the seven scenarios. To simplify these responses, we can consolidate this information into 3 different categories.

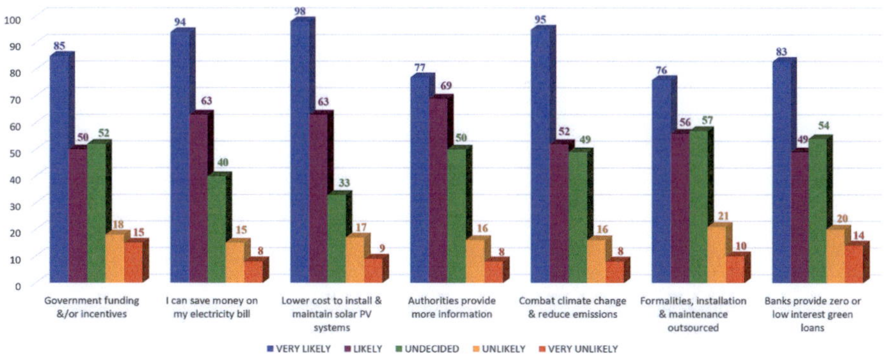

Fig. 9.11 Investment and utilization of solar energy. *Source* Created by researcher

It is evident from Table 9.4 that the majority of the respondents would consider investing and utilizing solar energy, if they were made aware of incentives, schemes or if they had sufficient information to consider.

Question 12:*Barriers/Challenges Impacting the Utilization of Solar Energy*

Various barriers were listed based on industry professionals, online research study, manufacturers, and potential customers. These challenges were presented to the respondents who listed the top three barriers that are presently hindering the expansion of the solar energy market in Dubai. These responses have been displayed in the pie chart (Fig. 9.12) shown above. As per they data generated and analyzed, the top three barriers restricting this industry growth is as follows:

- 23.18%—Lack of awareness about solar energy, its importance and benefits.
- 22.42%—High installation and maintenance costs.

Table 9.4 Likelihood of investing and using solar energy

Q	How likely am I to consider investing & using solar energy based on these possible scenarios?	LIKELY	%	UNDECIDED	%	UNLIKELY	%
1	Government funding &/or incentives	135	61.36%	52	23.64%	33	15.00%
2	I can save money on my electricity bill	157	71.36%	40	18.18%	23	10.45%
3	If the cost to install & maintain solar energy generation equipment is lowered	161	73.18%	33	15.00%	26	11.82%
4	Authorities provide more information about accessibility & availability	146	66.36%	50	22.73%	24	10.91%
5	To combat climate change & reduce greenhouse gas emissions	147	66.82%	49	22.27%	24	10.91%
6	A 3rd party undertakes all the government formalities, installation & maintenance work	132	60.00%	57	25.91%	31	14.09%
7	If banks can provide a zero or low interest loan to install Solar PV systems	132	60.00%	54	24.55%	34	15.45%

Source Created by researcher

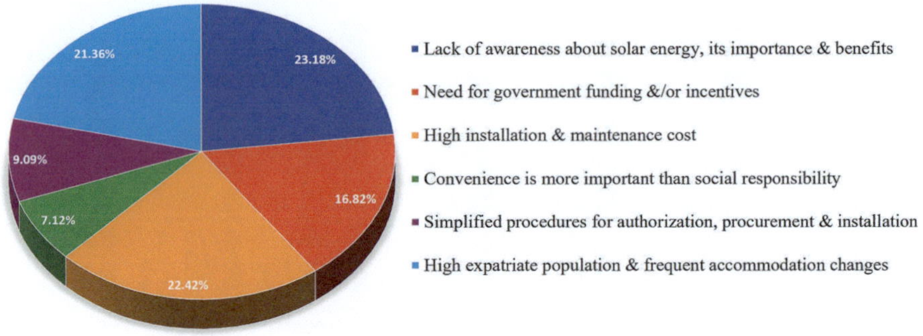

Fig. 9.12 Barriers/challenges in solar energy utilization. *Source* Created by researcher

- 21.36%—High expatriate population and frequent accommodation changes.

Together these 3 barriers make up for 66.97% of the responses and are critical to the successful implementation of solar energy utilization and our race toward sustainable energy management.

Statistical/Inferential Analysis

'Statistical Analysis' provides deeper insight into the relationships among various study aspects. Levene's T-Test and linear regression analysis were conducted to understand these relationships.

Levene's T-Test/Hypothesis Test

Understanding the mindset of 'business owners' and 'employees' is influenced by social awareness, current affairs, responsibilities, and capabilities. Levene's test for equality of variances analyzes variances between these groups. Using a 95% confidence interval, there is a 5% chance of error (alpha = 0.05). Two tests were conducted using SPSS with 34 responses from each category to ensure comparability (Table 9.5).

Table 9.5 Levene's test 1—knowledge and awareness

Group Statistics

	VAR00002	N	Mean	Std. Deviation	Std. Error Mean
VAR00001	Own	34	3.0000	1.07309	.18403
	Emp	34	2.9706	1.14111	.19570

Independent Samples Test

		Levene's Test for Equality of Variances		t-test for Equality of Means	
		F	Sig.	t	df
VAR00001	Equal variances assumed	.821	.368	.109	66
	Equal variances not assumed			.109	65.752

		t-test for Equality of Means			
		Significance			
		One-Sided p	Two-Sided p	Mean Difference	Std. Error Difference
VAR00001	Equal variances assumed	.457	.913	.02941	.26864
	Equal variances not assumed	.457	.913	.02941	.26864

		t-test for Equality of Means	
		95% Confidence Interval of the Difference	
		Lower	Upper
VAR00001	Equal variances assumed	-.50694	.56576
	Equal variances not assumed	-.50698	.56580

Source SPSS/Created by researcher

Levene's T-Test 1: Knowledge and Awareness of Photovoltaic Energy

Levene's T-Test 2: Investing and Utilizing Solar Energy

This test will analyze and determine the variance between 'Business Owners' (denoted by 'Own') and 'Employee' (denoted by 'Emp') in terms of investing and utilizing solar (PV) energy (Table 9.6).

Table 9.6 Levene's test—Investing and utilizing solar energy

Group Statistics

	VAR00002	N	Mean	Std. Deviation	Std. Error Mean
VAR00001	Own	34	3.8655	1.13468	.19460
	Emp	34	3.8866	.86058	.14759

Independent Samples Test

		Levene's Test for Equality of Variances		t-test for Equality of Means	
		F	Sig.	t	df
VAR00001	Equal variances assumed	2.105	.152	-.086	66
	Equal variances not assumed			-.086	61.526

		t-test for Equality of Means			
		Significance			
		One-Sided p	Two-Sided p	Mean Difference	Std. Error Difference
VAR00001	Equal variances assumed	.466	.932	-.02101	.24423
	Equal variances not assumed	.466	.932	-.02101	.24423

		t-test for Equality of Means	
		95% Confidence Interval of the Difference	
		Lower	Upper
VAR00001	Equal variances assumed	-.50864	.46662
	Equal variances not assumed	-.50930	.46728

Source SPSS/Created by researcher

Linear Regression

Linear regression is an analysis technique that is used to measure and compare the relation between dependent and independent variables. By carrying out this analysis, we can predict patterns between these variables simultaneously. In this case and with reference to solar energy, we will consider that 'saving money on electricity bills' would be the dependent variable, while various methods that could enable us to do so, will be considered as the independent variables. The calculations below are based on 220 responses for each of the variables on a 5-point Likert scale and carried out on SPSS software (Tables 9.7 and 9.8).

Table 9.7 Regression analysis—descriptive statistics

	Mean	Std. Deviation	N
Save money on my electricity bill	4.0000	1.10209	220
Government funding & / or incentives	3.7818	1.23408	220
All work undertaken by a 3rd Party	3.7591	1.15910	220
Banks provide zero or low interest loans	3.7591	1.22797	220
I need more information	3.8136	1.12538	220

Source Created by researcher via SPSS

Correlations

The correlation of a dependent variable is being tested against four other independent variables. Based on the analysis carried out, there is a strong positive correlation between all the variables. This implies that if the value of the dependent or independent variable increases, the other will also increase. The analysis also indicates that there is no multicollinearity between the independent variables which signifies that one independent variable cannot be predicted from another independent variable (Fig. 9.13).

Outliers Check

An outliers check has been carried out to identify if there are any extreme values compared to the other observations in the data set. From the above image, it is evident for all four independent variables, that there are no external points near the whiskers. Hence, there are no outliers, and the data quality is considered good for regression analysis.

Model Summary

Based on the model summary Table 9.9, the adjusted 'R' square value is 0.593. This value presents a score that 59.3% of the dependent variable is explained by the independent variables that have been considered for this analysis (Table 9.10).

Table 9.8 Regression analysis—correlation

		Save money on my electricity bill	Government funding & / or incentives	All work undertaken by a 3rd Party	Banks provide zero or low interest loans	I need more information
Pearson Correlation	Save money on my electricity bill	1.000	.685	.636	.584	.490
	Government funding & / or incentives	.685	1.000	.560	.492	.457
	All work undertaken by a 3rd Party	.636	.560	1.000	.649	.441
	Banks provide zero or low interest loans	.584	.492	.649	1.000	.304
	I need more information	.490	.457	.441	.304	1.000
Sig. (1-tailed)	Save money on my electricity bill	.	<.001	<.001	<.001	<.001
	Government funding & / or incentives	.000	.	.000	.000	.000
	All work undertaken by a 3rd Party	.000	.000	.	.000	.000
	Banks provide zero or low interest loans	.000	.000	.000	.	.000
	I need more information	.000	.000	.000	.000	.
N	Save money on my electricity bill	220	220	220	220	220
	Government funding & / or incentives	220	220	220	220	220
	All work undertaken by a 3rd Party	220	220	220	220	220
	Banks provide zero or low interest loans	220	220	220	220	220
	I need more information	220	220	220	220	220

Source Created by researcher via SPSS

Coefficients[a]

The coefficients measured through the analysis carried out are less than 0.05 which is the alpha value for 95% confidence. This result indicates that each of the independent variables has a statistically significant contribution to the dependent variable. The unstandardized coefficients indicate that for every 1 unit of change for each of these variables, we will see a corresponding increase in the monetary savings which is the dependent variable.

Linear regression determines the relation between two variables 'X' and 'Y' which are denoted by the equation '$Y = MX + C$'. In this equation, 'Y' is the dependent variable, 'M' is that amount which influences 'Y' and 'C' is the constant

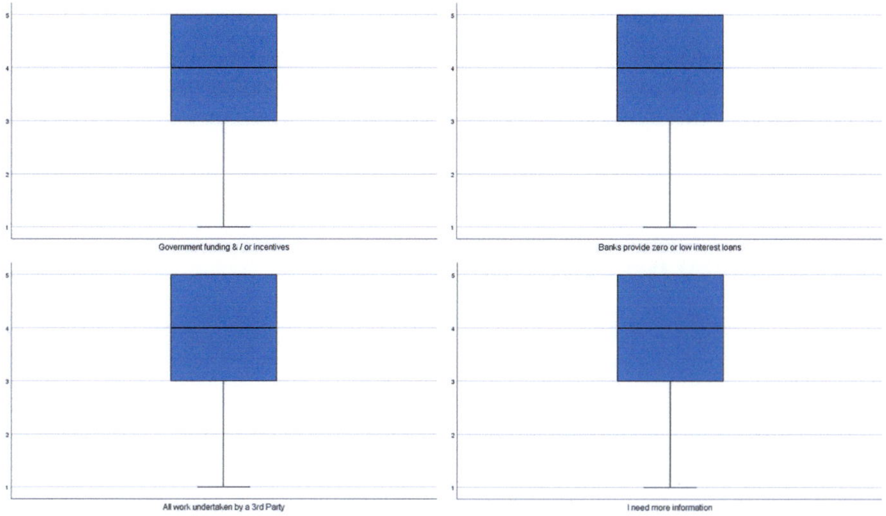

Fig. 9.13 Regression analysis—outliers check. *Source* Created by researcher via SPSS

Table 9.9 Regression analysis—model summary

Model	R	R Square	Adjusted R Square	Std. Error of the Estimate	Change Statistics				
					R Square Change	F Change	df1	df2	Sig. F Change
1	.775[a]	.601	.593	.70289	.601	80.85	4	215	<.001

a. Predictors: (Constant), I need more information, Banks provide zero or low interest loans, Government funding & / or incentives, All work undertaken by a 3rd Party

Source Created by researcher via SPSS

Table 9.10 Regression analysis—coefficients[a]

Model		Unstandardized Coefficients		Standardized Coefficients	t	Sig.	95.0% Confidence Interval for B		Correlations		
		B	Std. Error	Beta			Lower Bound	Upper Bound	Zero-order	Partial	Part
1	(Constant)	.642	.203		3.17	.002	.243	1.042			
	Government funding & / or incentives (X1)	.352	.049	.394	7.11	<.001	.254	.450	.685	.436	.307
	All work undertaken by a 3rd Party (X2)	.206	.059	.217	3.47	<.001	.089	.324	.636	.230	.150
	Banks provide zero or low interest loans (X3)	.182	.052	.203	3.50	<.001	.079	.284	.584	.232	.151
	I need more information (X4)	.149	.049	.152	3.03	.003	.052	.246	.490	.203	.131

a. Dependent Variable: Save money on my electricity bill

Source Created by researcher via SPSS

that represents any value that is unexplained. This equation is considered only if X and Y are correlated. In situations where there are multiple variables, the equation considered is $Y = M1X1 + M2X2 + MnXn + C$. Based on the equation above, and the coefficients generated from the linear regression analysis, this can be represented as $Y = (0.352 \times 1) + (0.206 \times 2) + (0.182 \times 3) + (0.149 \times 4) + 0.642$.

Conclusion and Recommendations

This research aimed to investigate the barriers to solar energy utilization in Dubai, assess current usage, and identify ways to overcome these barriers. The study found a positive consumer mindset toward solar energy.

- **Objective 1**: Barriers and knowledge gaps were identified through questions 6, 7, and 8, with successful results.
- **Objective 2**: The extent of PV grid utilization was assessed in questions 9 and 10, showing significant consumer association and understanding.
- **Objective 3**: Strategies to overcome barriers were addressed in questions 11 and 12, with positive responses indicating successful outcomes.

The study highlights the importance of addressing these barriers to contribute to the United Nations' Sustainability Development Goals. A significant knowledge gap exists, requiring attention from organizations and authorities to increase public awareness and utilization of solar energy. Respondents provided valuable insights for overcoming these barriers, emphasizing the potential environmental, business, and financial benefits.

Recommendations

This research has successfully addressed the objectives of this study. Yet we can conclude that there is a lot more that can be done to address global climate action and surely more research can be carried out to further dwell within the boundaries of this critical subject. To further enhance this research, the following recommendations are being suggested for consideration.

1. Authorities should engage consumers and highlight solar energy benefits through media and surveys to achieve UN's Sustainable Development Goals.
2. Simplify procedures for installing solar equipment and provide outreach materials to encourage investment.
3. Corporations should promote key information and product knowledge to boost industry growth.
4. Local manufacturing could reduce product costs and lead times.

5. Conduct a well-structured survey with a longer timeframe for more accurate data collection.
6. Perform a feasibility study on equipment manufacturing costs, lead times, and product life cycle to understand potential savings.

References

Ahmad, L., et al. (2020). Recent advances and applications of solar photovoltaics and thermal technologies. *Energy, 207*, 118254. https://doi.org/10.1016/j.energy.2020.118254

Basile, V., Loia, F., & Vona, R. (2021). Toward a sustainable decommissioning of offshore platforms in the oil and gas industry: A PESTLE analysis. *Sustainability, 13*(11), 6266. https://doi.org/10.3390/su13116266

Brummer, V. (2018). Community energy—Benefits and barriers: A comparative literature review of community energy in the UK, Germany and the USA, the benefits it provides for society and the barriers it faces. *Renewable and Sustainable Energy Reviews, 94*, 187–196. https://doi.org/10.1016/j.rser.2018.06.013

Chen, C., et al. (2021). Recent advances in solar energy full spectrum conversion and utilisation. *ES Energy & Environment*, 3–18. https://doi.org/10.30919/esee8c416

DEWA. (2022a). *Dubai—Solar Community [online], Dubai Electricity and Water Authority*. Accessed May 21, 2022. Available at: https://www.dewa.gov.ae/en/consumer/solar-community/shams-dubai

DEWA. (2022b). *Dubai—Sustainability reports [online], Dubai Electricity & Water Authority*. Accessed: May 21, 2022. Available at: https://www.dewa.gov.ae/en/consumer/Sustainability/sustainability-reports

Part II
Rating System in a Sustainable Environment

Chapter 10
Influencing Consumer Adoption of Sustainable Brands Using a New Framework for Sustainability Marketing

Jogitha Kanappaly, Nikhil Deshapande, Sandip Maiti, and Mihir Herlekar

Abstract This exploratory research investigates the discrepancy between consumer intentions and behaviors in the context of purchasing sustainable brands, providing a strategic model for marketing practitioners to mitigate this divergence. The study, involving a sample size of 124 participants, revealed that 48% actively seek sustainability certifications, while 80% exhibit behaviors consistent with sustainability. Despite these tendencies, a pronounced discrepancy in purchasing decisions remains, with 82% citing the lack of ethical practices as an impediment to sustainable purchases. This research introduces an innovative framework centered on 'Circularity Service' and other critical elements to address this gap. 'Circularity Service,' attracting 23% of consumer interest, emphasizes the significance of 'participatory behavior', advocating for consumer engagement in environmentally conscious activities. This approach advocates for brands to facilitate the translation of consumer's sustainable intentions into actionable behaviors and highlights the imperative role of consumer participation in making ecologically responsible decisions. The primary objective of this research is to pinpoint effective sustainability communication strategies that resonate with consumers, potentially leading to increased purchase behavior and long-term brand loyalty.

Keywords Intent-behavior-action gap · Sustainable shopping · Consumer behavior · Ethical branding · Circularity as a service · Participatory behavior

J. Kanappaly (✉) · N. Deshapande · S. Maiti
Tikao Technologies, Seawoods, Navi Mumbai, Maharashtra, India
e-mail: jogitha@tikao.in; jogithakjosey@gmail.com

J. Kanappaly · M. Herlekar
The Department of Environmental Science, The Institute of Science, Mumbai, Maharashtra, India
e-mail: tevs.mihir@iscm.ac.in

© The Author(s), under exclusive license to Springer Nature Switzerland AG 2026
J. Dsilva (ed.), *Redefining Sustainability*, https://doi.org/10.1007/978-3-031-88376-7_10

Introduction

This research paper investigates the intent-behavior-action gap in sustainable shopping, focusing on the divergence between consumers' eco-conscious aspirations and their actual purchasing behaviors. A comprehensive survey of 124 participants revealed that 48% actively seek out sustainability labels and 80% demonstrate sustainable habits, yet a significant gap in their purchasing decisions persists, predominantly influenced by the presence or absence of ethical practices in brands.

Drawing insights from seminal reports such as Deloitte's (2023) study 'Green Products Come of Age: Rising Expectations and Rewards for Sustainable Brands' (Deloitte, 2023) and Bain and Company's (2022) research, this study delves into factors influencing consumer purchase considerations and decisions. Furthermore, the research incorporates insights from McKinsey and Company's (2023) report on the growing prioritization of sustainability in consumer sectors (McKinsey & Company, 2023), highlighting the evolving consumer preferences toward ethically transparent brands.

The study aims to dissect and understand the dynamics driving consumer choices in sustainable shopping, investigating the roles of sustainability labels, ethical branding, and participatory practices such as 'Circularity Service'. This research provides a comprehensive analysis of how growing consumer awareness and expectations regarding sustainability influence shopping habits (Bain & Company, 2022).

Through this exploration, the paper seeks to contribute valuable insights to the academic and practical understanding of sustainable consumer behavior, offering strategic guidance for marketers and brands to align their practices with evolving consumer values and preferences, thereby bridging the gap between sustainable intentions and tangible actions (Deloitte, 2023; McKinsey & Company, 2023; Bain & Company, 2022).

Sustainability Brand Marketing Framework

Responding to key insights, our study introduces an innovative eight-pillar framework for sustainable marketing, emphasizing 'Circularity Service'. Drawing from industry research, including Bain & Company (2022) and Deloitte (2023), this framework seeks to enhance consumer engagement and surpass traditional marketing models. 'Circularity Service' plays a central role in involving consumers in sustainability and influencing brand perception and purchase decisions. Our methodology utilized robust quantitative analysis with a structured questionnaire to evaluate factors like sustainability label awareness, purchasing frequency, and shopping decision influences, with a focus on Circularity Service's impact on consumer behavior in the sustainable product sector.

Data analysis involved descriptive statistics to summarize findings clearly, highlighting key trends. As the research is in a pilot phase, it aims to establish a basic understanding of the data. Future phases will include inferential statistical methods for deeper data exploration and potential discovery of causal relationships.

Results

The majority of our survey's 124 respondents were from India, primarily married professionals aged 25 to 50. Gender-wise, 58% were male, 40% female, and 1.5% non-binary. The survey revealed that 48% of consumers check labels for sustainability, indicating health and environmental awareness, while 52% do not, suggesting a need for better consumer education on sustainability. Notably, label checking is more prevalent in food than fashion.

A significant 80% of respondents engage in sustainable habits, indicating a ready market for sustainability marketing. However, 20% are yet to adopt such practices, presenting an opportunity for effective marketing and education to increase sustainable behaviors. Additionally, 82% of respondents consider poor ethics a deal-breaker in supporting brands, emphasizing the importance of ethical considerations in consumer decisions.

Our survey screened 91% of participants as 'Sustainable Consumers', while 9% did not meet the criteria for sustainable purchasing habits. Furthermore, 59% are 'Occasional Ethical Shoppers', 22% are 'Guilt-Free Shoppers', and 19% are 'Consistent Conscientious' shoppers. This highlights a range of consumer behaviors toward ethical shopping. The survey also shows 24% of consumers favor energy-efficient appliances, with recycled paper goods and organic foods being popular choices. This trend presents an opportunity for marketers in sectors like sustainable fashion and eco-friendly cleaning supplies.

Our findings indicate that 48% of shoppers prioritize sustainability, while 33% prioritize cost, highlighting the need for competitively priced sustainable goods. Additionally, 'Circularity Service' is the most influential factor for 23% of respondents, followed by 'Certifications' (18%), 'Sustainable Packaging' (13%), and 'Clean Factory' processes (12%). Other factors like 'Transparency', 'Expression of Intent', 'Ethically Sourced', and 'Socially Responsible' practices also play a role in consumer decision-making.

In conclusion, consumer decisions on sustainable brands are influenced by factors such as pricing, accessibility, and credibility of sustainability claims. Our research, inspired by insights from Deloitte's May 2023 report, provides a structured approach for marketers to integrate elements like CSR reports, ESG disclosures, and other sustainability information into effective marketing campaigns. This strategy aligns with evolving consumer expectations and ensures impactful communication.

In the comprehensive paper, we delve into an in-depth discussion of each pillar, but for a brief overview, here are the names of the key pillars that constitute our framework: Expression of Intent, Ethical Sourcing, Verified Certifications, Social Responsibility, Clean Factory, Sustainable Packaging, Circularity Service, and Transparency and Traceability. These pillars represent the foundational elements of our sustainable marketing strategy, embodying the core principles and practices that are essential for aligning with contemporary consumer expectations and ethical standards.

Implications for Marketers

Marketers in sustainable shopping must emphasize trust and transparency to align consumer intentions with purchasing behaviors. By promoting 'Circularity Service' and participatory practices, they can engage consumers more deeply, fostering understanding and commitment to sustainable shopping. This engagement, crucial for transforming interest into action, requires campaigns that inform and empower ethical consumer choices.

Conclusion

The study highlights the importance of ethical considerations and consumer involvement in sustainable shopping. Effective sustainability promotion goes beyond traditional advertising, requiring holistic strategies that engage, educate, and empower consumers. These approaches enhance brand loyalty by resonating with consumer values and aspirations. Success for sustainable brands lies in authentically connecting with consumers, offering meaningful opportunities for positive change.

References

Bain & Company. (2022). *Meet Asia-Pacific's sustainable consumers.* [Infographic]. https://www.bain.com/insights/meet-asia-pacifics-sustainable-consumers-infographic/

Deloitte. (2023). *Global Marketing Trends 2020: Customer participation.* https://www.deloitte.com/global/en/our-thinking/insights/topics/marketing-sales/global-marketing-trends/2020/customer-participation.html

McKinsey & Company. (2023). *Prioritizing sustainability in the consumer sector.* https://www.mckinsey.com/industries/retail/our-insights/prioritizing-sustainability-in-the-consumer-sector

ResearchGate. (2023). *Influences of customer participation and customer brand engagement on brand loyalty.* https://www.researchgate.net/publication/305877806_Influences_of_customer_participation_and_customer_brand_engagement_on_brand_loyalty

Think with Google. (2023). *Sustainability marketing playbook.* https://www.thinkwithgoogle.com/future-of-marketing/digital-transformation/sustainability-marketing-playbook/

Chapter 11
From Green to Sustainable: A Holistic Evaluation of Social and Economic Dimensions of Green Building Rating Systems in Dubai

Zainab Rao, Mohd. Khalid Hassan, and Amir Khan Chauhan

Abstract The concept of sustainability is multifaceted, encompassing environmental, economic, and social dimensions, often referred to as the triple bottom line. This comprehensive approach is visually represented by interlinked circles, highlighting the inherent interconnectedness of these dimensions. Achieving sustainability in construction necessitates addressing environmental, economic, and social aspects. While environmental dimension has received considerable attention in green building assessments, the socio-economic aspects of sustainable development have been overlooked (Atanda & Öztürk, 2020). This research places a particular emphasis on understanding the social implications of green building rating systems. The primary goal is to explore and develop a comparative analysis between five well-established green building rating tools which are majorly used in UAE such as international (LEED and BREEAM), regional (GSAS), and national (Estidama, Al Sa'fat), with an overarching objective of enhancing community well-being. This paper attempts to provide a brief global discussion of the social-economic aspects of GBRSs, with a particular focus on the local green building system, Al Sa'fat. The goal of this study is to find viable recommendations that may be made to promote sustainability at the local level. Through a comprehensive analysis of current green rating systems and their effects on socioeconomic factors, the study aims to offer insightful recommendations for improving approaches to sustainable community development. The study concludes with recommendations for policy-level initiatives that stress the inclusion of social-economic factors in the evaluation of sustainable buildings and communities. This methodical approach offers helpful suggestions for enhancing the socio-economic impact of green building initiatives, which adds to the larger conversation on sustainable development.

Z. Rao (✉) · Mohd. K. Hassan
Department of Architecture, Aligarh Muslim University, Aligarh, India
e-mail: gg4515@myamu.ac.in

A. K. Chauhan
Principal Transport Planner, Zutari Engineering Consultants, Dubai, UAE

© The Author(s), under exclusive license to Springer Nature Switzerland AG 2026

J. Dsilva (ed.), *Redefining Sustainability*, https://doi.org/10.1007/978-3-031-88376-7_11

Keywords Green building rating systems (GBRSs) · Socio-economic criterion · Sustainable neighborhoods

Introduction

Green building assessment tools are now well-established in the pursuit of sustainable development within the built environment. This has sparked considerable debate and has been the subject of research from various perspectives. These tools generally fall into two primary categories: criteria-based tools and life cycle assessment tools. The majority of research efforts have been directed toward criteria-based assessment tools, leading to the development of various systems such as CASBEE, BREEAM, Green Star, LEED, among others (Atanda & Öztürk, 2020). These technologies have been developed in response to the growing need to address global concerns. The adoption of ecologically and socially friendly practices in this sector has been recognized as vital for decades, as the built environment is one of the key contributors to socio-ecological challenges, above all being climate change (Marchi et al., 2021). The building industry must become more sustainable due to its significant influence on the socio-economic and environmental issues that are receiving more and more attention globally. All parties involved must voluntarily work to design, build, operate, and maintain buildings while taking a comprehensive approach to sustainability, in addition to laws and policies being enforced. This necessitates having a thorough understanding of both potential mitigating measures and the impact of construction characteristics on the triple bottom line (planet, people, and profit) (Marchi et al., 2021).

Despite the abundance of professional perspectives and research efforts, a conventional solution has been widely embraced. Unfortunately, the focus on environmental criteria has dominated the landscape of most building assessment tools, overshadowing the importance of addressing social criteria (Atanda & Öztürk, 2020). As a result, sustainability evaluation has been acknowledged as a critical tool for achieving this goal, and green building rating systems (GBRSs) have become an important instrument for evaluating and directing the entire building process to be more environmentally friendly (Ding, 2008; Marchi et al., 2021).

Methodology

According to the World Green Building Council (2015), out of 71 green building council members, 27 councils were identified as established green building councils. Therefore, these 27 established green building councils were classified under the main geographic regions and based on the extent of usage; widely used green building rating tools were selected for the study. Five well-established green building rating tools which are majorly used in UAE were further selected such as international

(LEED and BREEAM), regional (GSAS), and national (Estidama), Al Sa'fat. These GBRSs were studied to identify the environmental, economic, and social parameters considered for the rating of a building and a neighborhood. A thorough literature study has been conducted for a tangible and intangible comparative analysis of these GBRSs on the basis of environmental, economic, and social aspects. The investigation of green building assessment tools focused on the social criteria of sustainable development. The study incorporated available literature as a key element for discussion, providing a foundation to substantiate the research objective. Initially, 36 research papers from QI and Q2 journals are chosen; of these, 25 research papers—including one from a government guidelines—are considered very important in the context of the UAE and the rest of the world. Selected papers include critical reviews of the literature, bibliometric reviews of the literature, and comparisons between different GBRSs. Different databases, including Science Direct, Web of Science, Scopus and Google Scholar, are used to choose the papers. A number of journals are cited as the foundation for this research, including knowledge-based systems, Journal of Building Engineering, Journal of Cleaner Production, Procedia-Social and Behavioral Sciences, Architectural Science Review, Building and Environment, IOP Conference Series: Earth and Environmental Science, Energy and Buildings, Energy Strategy Reviews, Environment, Development and Sustainability, Sustainability, Journal of Professional Issues in Engineering Education and Practice, Energy Policy, and Dubai municipality guidelines. To systematically categorize each system credit under the umbrella of sustainable criteria, a mixed approach was employed as a strategic research methodology. This approach utilized a bottom-up methodology to conduct a comprehensive review of green building assessment tools across UAE (LEED, BREEAM, GSAS, ESTIDMA, Al Sa'fat). In this study, the adoption of social criteria categories was grounded in the analysis of relevant literature, facilitating a refined and targeted exploration of the socio-economic dimensions within the chosen context.

Green Building Rating Systems

To put it briefly, green building rating systems (GBRSs) are voluntary, third-party, market-driven standards that use multi-criteria assessments to gauge a building's sustainability and promote the use of ecologically, socially, and economically sound building (or neighborhood) design, construction, and operation practices (Marchi et al., 2021). Green building and rating system are two distinct components that are included in one concept, therefore defining the two independently may aid in providing a good definition for the entire. Initially, the term "green building" encompasses a wide range of topics and is a complex idea. As per the definition provided by the World Green Building Council (World GBC), a building is defined as one that can have good effects on the environment and climate while also reducing or eliminating negative effects through its design, construction, or operation. Natural resources are preserved, and our quality of life is enhanced by green construction (Marchi et al., 2021).

BREEAM, LEED, GSAS, ESTIDMA, and Al Sa'fat

The Building Research Establishment (BRE) in the UK created and runs BREEAM, which is regarded as the first green building grading system in the world (Alyami & Rezgui, 2012; Chen et al., 2015; Doan et al., 2017). It was first released to the public in 1990 and had its first revision in 1993 to evaluate offices (Doan et al., 2017; Lee, 2013). It is commonly acknowledged that BREEAM had a significant influence on practically all subsequent major green grading systems, including LEED, Green Star, and CASBEE (Doan et al., 2017). The flexibility of BREEAM makes it a popular choice. It allows application in international buildings in addition to evaluating local codes and conditions. BREEAM International for New Construction 2016 is the most recent of the international versions that have been made available for project certification across the globe. A percentage of success over the entire number of points available is used to convey the system's evaluation: 30% for pass classification, 45% for good, 55% for very good, 70% for excellent, and 85% for outstanding. Management, Health & Wellbeing, Energy, Transport, Water, Materials, Waste, Land Use & Ecology, Pollution, and Innovation are the categories for new construction projects (Awadh, 2017; BREEAM-International-New-Construction-2016, n.d.).

Launched in 1998 by the US Green Building Council (USGBC), the LEED grading system stands for Leadership in Energy and Environmental Design. The most well-liked and extensively applied green building grading system is LEED. Based on the point criteria met, there are four certification levels for LEED version 4 for New Construction (NC), which was last released in 2014. These include Certified (40–49 points), Silver (50–59 points), Gold (60–79 points), and Platinum (80 points and above). Seven assessment categories are available for a total of 126 points (Awadh, 2017; LEED v4 for BUILDING DESIGN AND CONSTRUCTION Includes: LEED BD+C: New Construction LEED BD+C: Core and Shell LEED BD+C: Schools LEED BD+C: Retail LEED BD+C: Data Centers LEED BD+C: Warehouses and Distribution Centers LEED BD+C: Hospitality LEED BD+C: Healthcare, 2019).

In 2009, the Gulf Organization of Research and Development (GORD) in Qatar launched the Integrated and Performance-Based Green Building Assessment Rating System (GSAS). Best practices from the most well-known global rating systems, such as BREEAM (United Kingdom), LEED (United States), GREEN GLOBES (Canada), CEPAS (Hong Kong), CASBEE (Japan), and the International SBTOOL, served as the basis for the creation of the Global Sustainability Assessment System (GSAS). There are eight GSAS categories, and certification in one to six stars is possible (Awadh, 2017; Issue 2, 2019).

Upon closer inspection, Estidama was primarily created with BREEAM and LEED components, tailoring the system to the requirements and environment of the region (Awadh, 2017; Elgendy, 2010). The Abu Dhabi Urban Planning Council

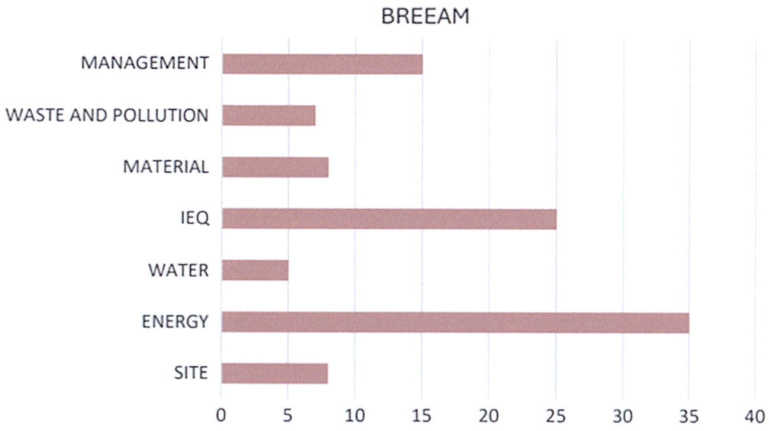

Fig. 11.1 Analysis of the assessment criteria of BREEAM (Alhamlawi et al., 2021)

(UPC) introduced the Estidama Pearl Rating System in 2010. The Pearl Building Rating System (PBRS) offers the following five certification levels: 1 Pearl, which just requires prerequisites; 2 Pearl, which requires requirements plus 60 points; 3 Pearl, which requires prerequisites plus 85 points; 4 Pearl, which requires prerequisites plus 115 points; and 5 Pearl, which requires prerequisites plus 140 points. The PBRS offers eight categories with a total of 180 points that can be earned (Awadh, 2017; The Pearl Rating System for Estidama Building Rating System Design & Construction, 2010).

The Dubai Municipality initially issued Al Sa'fat in 2016, which is the plural of the Arabic word for palm tree leaves and covers a range of construction kinds, including public, commercial, residential, and industrial buildings. Al Sa'fat is not a point-based system, in contrast to other GBRSs like LEED. When all requirements for one of the four certification levels—Bronze, Silver, Gold, or Platinum—are met, a building is certified. Ecology and Planning, Building Vitality, Resource Effectiveness for Energy, Resource Effectiveness for Water, and Resource Effectiveness for Materials and Waste are the five areas into which the certification requirements are divided (Alhamlawi et al., 2021; Municipality, n.d.) (Figs. 11.1, 11.2, 11.3, 11.4, 11.5, 11.6, 11.7, 11.8, and 11.9).

Limitations and Observations

The study's scope is restricted to evaluating five significant GBRSs in Dubai based on their social, environmental, and economic dimensions—a triple bottom line that also applies to GBRSs according to previous research. It has been noted that there are not

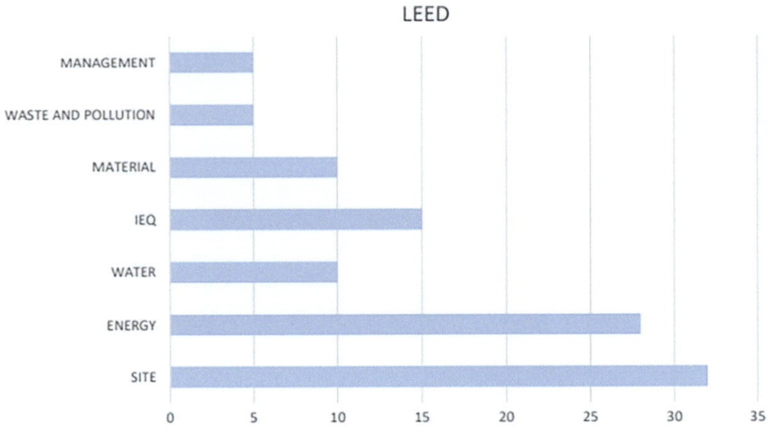

Fig. 11.2 Analysis of the assessment criteria of LEED (Alhamlawi et al., 2021)

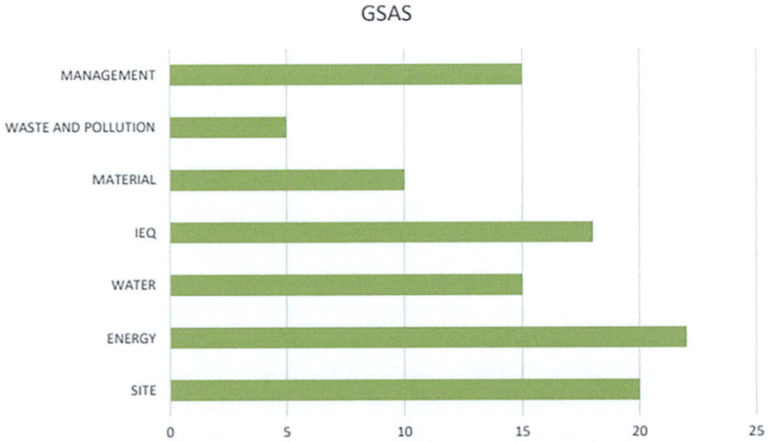

Fig. 11.3 Analysis of assessment criteria of GSAS (Alhamlawi et al., 2021)

many socioeconomic studies on green building rating systems accessible. Furthermore, the comparison analysis's parameters for social, economic, and environmental factors vary, which appears to be biased based on the literature available.

Findings and Suggestions

The research aims to propose recommendations for enhancing the inclusivity of green building rating systems (GBRSs) and establishing a balance among environmental, economic, and social criteria. In order to conduct a fair comparison examination of the

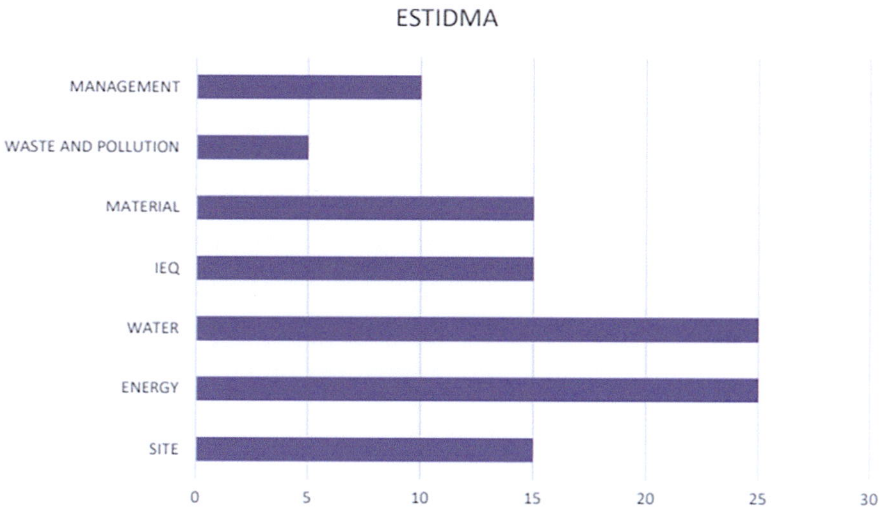

Fig. 11.4 Analysis of assessment criteria of ESTIDAMA (Alhamlawi et al., 2021)

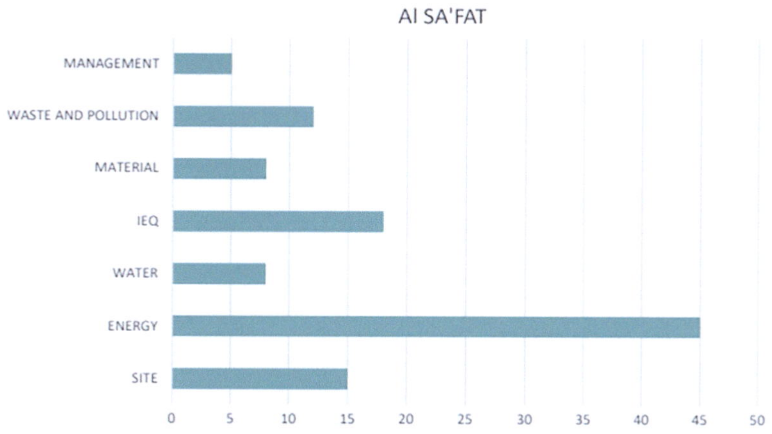

Fig. 11.5 Analysis of assessment criteria of Al Sa'fat (Alhamlawi et al., 2021)

green building rating systems, the World Green Building Council should first explicitly identify environmental, social, and economic characteristics. Secondly, green communities should also be taken into consideration and possibly made required, in addition to green buildings. Assessing a building separately could not have the same impact as taking the neighborhood as a whole. The implementation of a comprehensive and inclusive framework is necessary for sustainable growth. Thirdly, according to (Atanda & Öztürk, 2020), no GBRS has evidenced post-validation. Fourthly, as political sustainability is the key factor that will make it happen, it must be taken into account. These interventions are poised to make a meaningful contribution toward

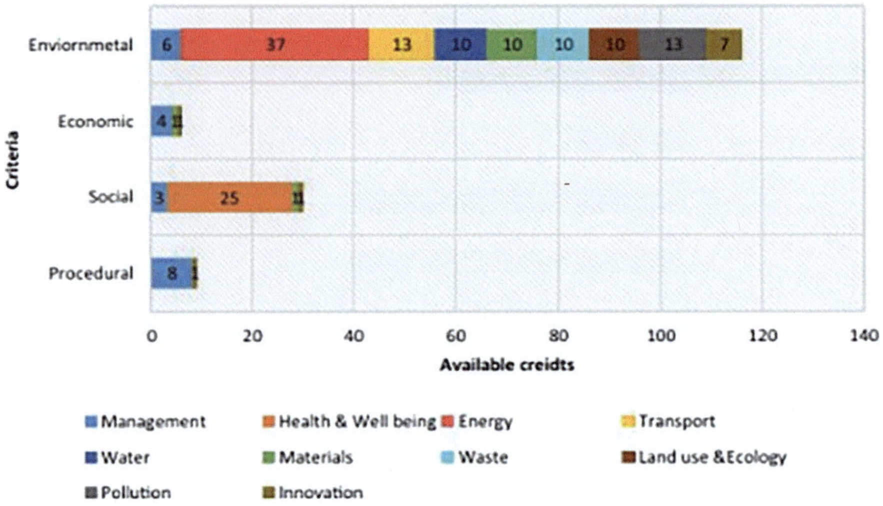

Fig. 11.6 Comparison of environmental, economic, and social aspects BREEAM (Awadh, 2017)

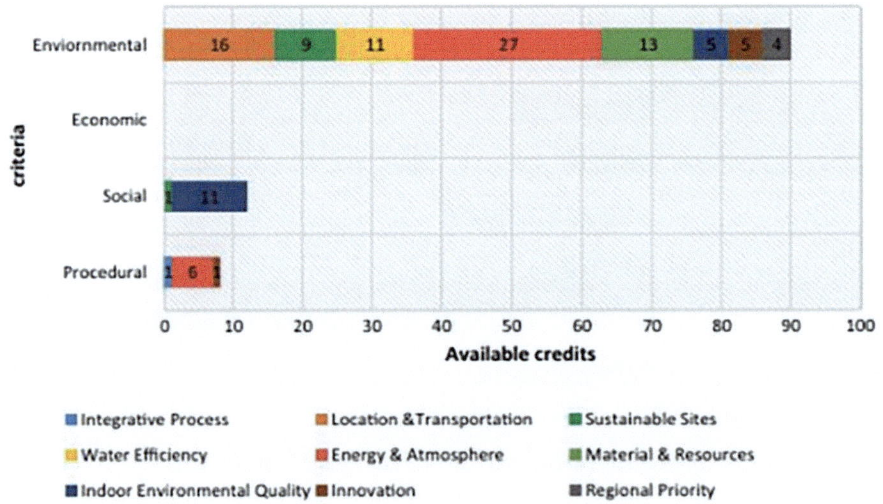

Fig. 11.7 Comparison of environmental, economic, and social aspects LEED (Awadh, 2017)

achieving Sustainable Development Goal 11 (Sustainable Cities and Communities), particularly targeting 11.3 of SDG 11.

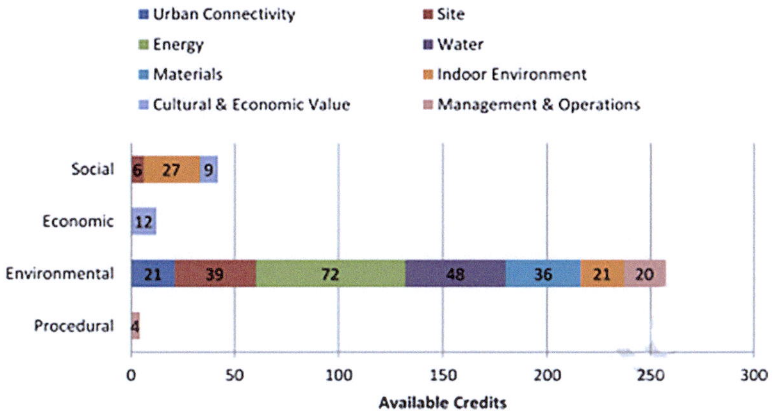

Fig. 11.8 Comparison of environmental, economic, and social aspects GSAS (Awadh, 2017)

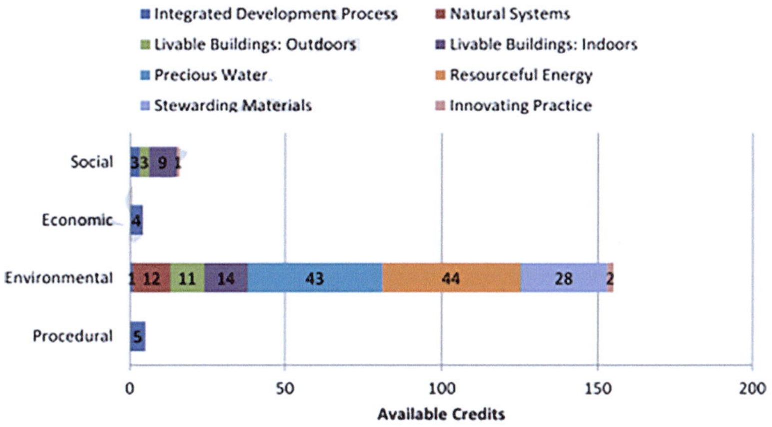

Fig. 11.9 Comparison of environmental, economic, and social aspects ESTIDMA (Awadh, 2017)

Conclusions

For development to be environmentally, economically, and socially sustainable, it is important to make an equilibrium between them. When evaluating sustainability, the human-centric approach should be taken into account. In order to achieve community and social inclusion, GBRSs should be modified. A framework for the examination of sustainable buildings must be created that takes into account social, economic, and environmental issues in equal measure and applies a micro-to-macroscopic (from building to neighborhood) approach. It is important to distinguish between the terms "green" and "sustainable" rather than using them interchangeably. To be clear, becoming green is only a component of sustainability (Doan et al., 2017; Runde & Thoyre, 2010; Wong & Zhou, 2015; Yanarella et al., 2009).

References

Alhamlawi, F., Alaifan, B., & Azar, E. (2021). A comprehensive assessment of Dubai's green building rating system: Al Sa'fat. *Energy Policy, 157*. https://doi.org/10.1016/j.enpol.2021.112503

Alyami, S.H., & Rezgui, Y. (2012). Sustainable building assessment tool development approach. *Sustainable Cities and Society*, *5*(1), 52–62. https://doi.org/10.1016/j.scs.2012.05.004

Atanda, J. O., & Öztürk, A. (2020). Social criteria of sustainable development in relation to green building assessment tools. *Environment, Development and Sustainability*, *22*(1), 61–87. https://doi.org/10.1007/s10668-018-0184-1

Awadh, O. (2017). Sustainability and green building rating systems: LEED, BREEAM, GSAS and Estidama critical analysis. *Journal of Building Engineering, 11*, 25–29. https://doi.org/10.1016/j.jobe.2017.03.010

BREEAM-International-New-Construction-2016. (n.d.).

Chen, X., Yang, H., & Lu, L. (2015). A comprehensive review on passive design approaches in green building rating tools. In *Renewable and sustainable energy reviews* (Vol. 50, pp. 1425–1436). Elsevier Ltd. https://doi.org/10.1016/j.rser.2015.06.003

Ding, G. K. C. (2008). Sustainable construction—The role of environmental assessment tools. *Journal of Environmental Management, 86*(3), 451–464. https://doi.org/10.1016/j.jenvman.2006.12.025

Doan, D. T., Ghaffarianhoseini, A., Naismith, N., Zhang, T., Ghaffarianhoseini, A., & Tookey, J. (2017). A critical comparison of green building rating systems. In *Building and Environment* (Vol. 123, pp. 243–260). Elsevier Ltd. https://doi.org/10.1016/j.buildenv.2017.07.007

Elgendy K. (2010). Comparing Estidama's pearls rating system to LEED and BREEAM. http://www.carboun.com/sustainable-urbanism/comparing-estidama%E2%80%99s-pearls rating-method-to-leed-and-breeam. http://www.carboun.com

Illankoon, I. M. C. S., Tam, V. W. Y., & Le, K. N. (2017). Environmental, economic, and social parameters in international green building rating tools. *Journal of Professional Issues in Engineering Education and Practice, 143*(2). https://doi.org/10.1061/(ASCE)EI.1943-5541.0000313. www.gord.qa

Lee, W. L. (2013). A comprehensive review of metrics of building environmental assessment schemes. *Energy and Buildings*, *62*, 403–413. https://doi.org/10.1016/j.enbuild.2013.03.014

LEED v4 for BUILDING DESIGN AND CONSTRUCTION Includes: LEED BD+C: New Construction LEED BD+C: Core and Shell LEED BD+C: Schools LEED BD+C: Retail LEED BD+C: Data Centers LEED BD+C: Warehouses and Distribution Centers LEED BD+C: Hospitality LEED BD+C: Healthcare. (2019).

Marchi, L., Antonini, E., & Politi, S. (2021). Green building rating systems (GBRSs). *Encyclopedia, 1*(4), 998–1009. https://doi.org/10.3390/encyclopedia1040076

Municipality, D. (n.d.). *Dubai green building system.*

Runde, T., & Thoyre, S. (2010). Integrating sustainability and green building into the appraisal process. *Journal of Sustainable Real Estate.*

The Pearl Rating System for Estidama Building Rating System Design & Construction. (2010). www.upc.gov.ae

Wong, J. K. W., & Zhou, J. (2015). Enhancing environmental sustainability over building life cycles through green BIM: A review. In *Automation in construction* (Vol. 57, pp. 156–165). Elsevier. https://doi.org/10.1016/j.autcon.2015.06.003

Yanarella, E. J., Levine, R. S., & Lancaster, R. W. (2009). Green versus sustainability. *Sustainability, 2*(5), 296–302. https://doi.org/10.1089/SUS.2009.9838

Chapter 12
Crafting Tailored Sustainability Standards for the UAE: A Comparative Analysis and Strategic Framework

Sameera Fernandes, M. S. Sheeja, and Swetha Parivara

Abstract In the face of unique environmental challenges and rapid urban expansion, the United Arab Emirates (UAE) stands at a crucial juncture regarding its approach to sustainability. This study aims to critically analyze prevailing global sustainability rating systems, assessing their applicability and limitations within the UAE's distinct context. The primary objectives include identifying gaps in these global frameworks when applied to the UAE's specific socio-economic and environmental conditions, and proposing a bespoke sustainability framework, thus contributing to the ongoing discourse on regional adaptation of global sustainability standards. Employing a qualitative research methodology, the study extensively reviews existing literature on global sustainability standards, complemented by a series of case studies focusing on their implementation within the UAE. This approach facilitates a nuanced comparative analysis, allowing for a detailed examination of the suitability and adaptability of these standards to the UAE's unique challenges. Key standards analyzed include the Global Reporting Initiative (GRI), the Sustainability Accounting Standards Board (SASB), among others. Insights from UAE-based sustainability experts and stakeholders further enrich the analysis, offering a local perspective on global standards. The study finds that while standards like GRI and SASB provide comprehensive guidelines for sustainability reporting, they do not sufficiently address specific UAE challenges such as extreme climatic conditions and cultural nuances. Similarly, frameworks like TCFD and CDP, with their focus on climate-related financial disclosures, demonstrate a limited scope in capturing the socio-cultural intricacies of the UAE. Evaluating the implementation of these standards in the UAE uncovers successes, particularly in energy and waste management. However, significant gaps are noted in integrating local cultural values and traditional practices into

S. Fernandes (✉) · M. S. Sheeja · S. Parivara
Garden City University & Century Financial, Bengaluru, India
e-mail: Sameera.fernandes@gmail.com

M. S. Sheeja
e-mail: sheeja.ms@gardencity.university

S. Parivara
e-mail: swetha.parivara@gardencity.university

© The Author(s), under exclusive license to Springer Nature Switzerland AG 2026
J. Dsilva (ed.), *Redefining Sustainability*, https://doi.org/10.1007/978-3-031-88376-7_12

modern sustainability frameworks. The research highlights crucial areas for adaptation, including enhanced focus on water conservation, societal responsibility, and cultural relevance. It suggests the potential for a hybrid model that blends global best practices with specific local needs.

Keywords Sustainability standards · Environmental adaptation · UAE-specific framework · Global–local integration · Renewable energy solutions

Introduction

Scope and Objective

The rapid urban and economic growth of the United Arab Emirates (UAE) presents unique environmental challenges, necessitating a tailored approach to sustainability. This research paper aims to bridge the gap between global sustainability standards and the specific needs of the UAE, by developing a customized sustainability framework. The primary objective is to understand how global standards can be adapted and integrated with local environmental, cultural, and socio-economic factors to create a comprehensive and applicable sustainability model for the UAE.

Review of Pertinent Literature

Existing literature on sustainability primarily focuses on global standards such as the Global Reporting Initiative (GRI) (GRI, 2021), the Sustainability Accounting Standards Board (SASB) (SASB, 2020), and the Task Force on Climate-related Financial Disclosures (TCFD) (TCFD, 2021). These standards, while effective in a global context, often overlook the unique challenges faced by regions like the UAE, including extreme climatic conditions, water scarcity, and rapid urbanization (Al-Saidi & Saliba, 2019; Kozak, 2022). Literature on region-specific sustainability practices, especially in the context of the Middle East, is relatively sparse (Zaabi et al., 2020). This research aims to fill this gap by drawing insights from global practices and aligning them with local needs.

Methods

The methodology adopted for this research includes a qualitative analysis of existing global sustainability standards and their application in various contexts, emphasizing the UAE (Bryman, 2016). Comparative analysis techniques are used to evaluate the effectiveness of these standards in addressing the specific challenges faced by the UAE (Creswell & Creswell, 2018). The study also incorporates insights from industry experts, policymakers, and academic literature to provide a comprehensive understanding of the subject (Jabbour et al., 2020).

Overview of Main Research Results

The research reveals that while global sustainability standards offer a robust foundation, their direct application in the UAE context is limited due to region-specific challenges (Khan et al., 2021). Key findings include the identification of gaps in current practices, particularly in areas like water conservation, energy efficiency, and cultural integration (Darmoul et al., 2019). The proposed framework suggests a hybrid approach combining global best practices with tailored strategies to address these gaps. This includes industry-specific guidelines, emphasis on renewable energy, innovative water management, and a strong focus on public–private partnerships and community engagement (Sarkis et al., 2020; Hammad & Akbar, 2021).

Methodology

Research Question and Relevance

The central research question of this study is: "How can global sustainability standards be effectively adapted to meet the unique environmental, cultural, and socioeconomic needs of the United Arab Emirates (UAE)?" This question is grounded in the recognition that while global sustainability standards provide a comprehensive framework (GRI, 2021; SASB, 2020; TCFD, 2021), their applicability in diverse regional contexts, such as that of the UAE, often requires significant adaptation (Khan et al., 2021; Kozak, 2022).

The relevance of this question is twofold: theoretically, it contributes to the literature on sustainability by providing insights into the adaptation of global frameworks to local contexts (Al-Saidi & Saliba, 2019; Jabbour et al., 2020); practically, it offers a guideline for policymakers and industry leaders in the UAE and similar regions to develop more effective and relevant sustainability strategies (Sarkis et al., 2020; Hammad & Akbar, 2021).

The research employs a multi-faceted framework comprising literature reviews, comparative analyses, expert consultations, and case studies to explore the adaptation of global sustainability standards to the UAE's context (Creswell & Creswell, 2018; Bryman, 2016).

The research combines a literature review of global sustainability standards with a comparative analysis tailored to the UAE's unique context, complemented by expert consultations and case studies to bridge theoretical frameworks with practical applications and insights (Zaabi et al., 2020; Darmoul et al., 2019).

Relevance to Theory and Practice

The methodology is designed to ensure that the research is relevant both to academic theory and practical application. Academically, it contributes to the growing body of literature on localized sustainability practices and the adaptation of global frameworks (Jabbour et al., 2020; Khan et al., 2021). Practically, the findings and recommendations are intended to guide policymakers and industry leaders in the UAE and similar regions in developing and implementing more effective sustainability strategies (Sarkis et al., 2020; Hammad & Akbar, 2021). This dual relevance ensures that the research has a tangible impact, advancing the field of sustainability and aiding in the development of more sustainable practices in the UAE and beyond.

Findings: Proposed Framework for the UAE

The proposed framework for tailoring sustainability standards specifically for the United Arab Emirates (UAE) is designed to address the unique environmental, cultural, and socio-economic aspects of the region. Here are the key elements of the framework:

1. **Integration of Local Environmental Concerns**: Addressing specific environmental challenges like water scarcity, high temperatures, and desertification (Al-Saidi & Saliba, 2019; Kozak, 2022).
2. **Cultural and Historical Preservation**: Incorporating traditional practices and heritage conservation into sustainability initiatives (Darmoul et al., 2019; Hammad & Akbar, 2021).
3. **Industry-Specific Sustainability Guidelines**: Developing tailored guidelines for key industries such as oil and gas, construction, and tourism (Khan et al., 2021; Zaabi et al., 2020).
4. **Emphasis on Renewable Energy**: Promoting the use of renewable energy sources, especially solar energy, to reduce dependence on fossil fuels (Sarkis et al., 2020; Jabbour et al., 2020).
5. **Innovative Water Management Solutions**: Implementing advanced technologies for water conservation and desalination, critical in the arid climate of the UAE (Al-Saidi & Saliba, 2019; Darmoul et al., 2019).
6. **Sustainable Urban Development**: Encouraging green building practices and sustainable urban planning to manage the rapid urbanization in the region (Kozak, 2022; Khan et al., 2021).

7. **Waste Reduction and Recycling**: Implementing strategies for waste minimization, recycling, and sustainable waste management (Sarkis et al., 2020; Hammad & Akbar, 2021).
8. **Public–Private Partnerships (PPPs)**: Fostering collaborations between government entities and private sectors for sustainability projects (Hammad & Akbar, 2021).
9. **Community Engagement and Education**: Raising public awareness and involvement in sustainability practices through education and community programs (Jabbour et al., 2020; Zaabi et al., 2020).
10. **Monitoring and Continuous Improvement**: Establishing robust mechanisms for monitoring sustainability metrics and continuously updating standards based on new research and technologies (GRI, 2021; SASB, 2020; TCFD, 2021).

This framework provides a comprehensive approach to developing and implementing sustainability standards that are not only effective in the unique context of the UAE but also set a precedent for other regions with similar environmental and socio-cultural landscapes.

Implications and Conference Relevance

This research on developing tailored sustainability standards for the UAE offers extensive implications in policymaking, industry practices, societal impacts, and global sustainability discourse.

Policymaking: It advocates for region-specific policies addressing the UAE's unique environmental, cultural, and economic contexts, suggesting refined regulatory frameworks and incentivization strategies to encourage sustainable practices (Khan et al., 2021; Al-Saidi & Saliba, 2019).

Industry Practices: It emphasizes the adoption of specific sustainability guidelines, innovation, and technology for efficient resource management, and the alignment of supply chains with sustainability standards (Sarkis et al., 2020; Jabbour et al., 2020).

Societal Impacts: It promotes community engagement, cultural integration, and addresses local environmental challenges to improve living conditions (Zaabi et al., 2020; Hammad & Akbar, 2021).

Global Sustainability Discourse: The UAE's approach serves as a model for similar regions, contributing to the sustainability dialogue and fostering international collaboration in integrating local considerations into broader sustainability frameworks (GRI, 2021; SASB, 2020; TCFD, 2021).

Conclusions

This study highlights the critical need for regionally tailored sustainability standards, especially in unique settings like the UAE. By advocating for a region-specific approach, the UAE can effectively address its distinct sustainability challenges while contributing to global sustainability efforts (Al-Saidi & Saliba, 2019; Khan et al., 2021). The research findings and the proposed framework offer substantial contributions to sustainability literature, particularly in the context of adapting and refining global standards to align with regional specificities (Jabbour et al., 2020; Sarkis et al., 2020).

This research has delved into the critical task of adapting global sustainability standards to the unique context of the United Arab Emirates (UAE), addressing its distinct environmental, cultural, and socio-economic challenges. The study has led to several key principles and generalizations:

1. **Adaptation of Global Standards**: Global sustainability frameworks such as GRI, SASB, and TCFD provide a robust foundation but require significant adaptation to be fully effective in the UAE (GRI, 2021; SASB, 2020; TCFD, 2021). This adaptation is necessary to address the UAE's specific challenges like extreme climate, water scarcity, and rapid urbanization (Zaabi et al., 2020; Kozak, 2022).
2. **Local Specificity in Policy and Practice**: The research underscores the importance of developing region-specific policies and practices that are deeply rooted in the local context of the UAE (Khan et al., 2021; Hammad & Akbar, 2021). These should prioritize local environmental concerns, cultural values, and the nation's developmental pattern.
3. **Collaborative Approach**: The study highlights the need for a collaborative approach involving government bodies, private sectors, and communities in developing and implementing sustainability strategies tailored to the UAE (Jabbour et al., 2020; Darmoul et al., 2019).

However, the work is not without its limitations. The dynamic nature of both global sustainability standards and the UAE's evolving landscape means that continual monitoring and adaptation of the frameworks are necessary (Sarkis et al., 2020). Moreover, the balance between maintaining global best practices and integrating local specifics poses a continuous challenge (Kozak, 2022).

The theoretical implications of this research contribute to the broader discourse on sustainability, particularly in the context of adapting global practices to local environments (Al-Saidi & Saliba, 2019; Khan et al., 2021). Practically, it offers actionable insights for policymakers, corporate leaders, and sustainability practitioners in the UAE and similar regions (Hammad & Akbar, 2021).

Practical Recommendations

The practical recommendations from this study include:

1. **Policy Reform**: Encouragement of policy reforms that align with the specific needs of the UAE, including incentives for sustainable practices and regulations that support industry-specific sustainability adaptations (Khan et al., 2021; Jabbour et al., 2020).
2. **Corporate Sustainability Integration**: Guidance for corporations in the UAE to integrate sustainability into their core operations, encouraging innovation and the adoption of cutting-edge technologies for sustainable development (Sarkis et al., 2020; Darmoul et al., 2019).
3. **Community Engagement and Education**: Emphasis on the importance of community engagement and public education in sustainability efforts, fostering a culture of sustainability across various sectors of society (Hammad & Akbar, 2021; Zaabi et al., 2020).

References

Bryman, A. (2016). *Social research methods* (5th ed.). Oxford University Press.

Creswell, J. W., & Creswell, J. D. (2018). *Research design: Qualitative, quantitative, and mixed methods approaches* (5th ed.). SAGE Publications.

Darmoul, S., El Mhamedi, A., & Lefebvre, B. (2019). A comprehensive framework for sustainable performance assessment in supply chains. *Journal of Cleaner Production, 233*, 1289–1303. https://doi.org/10.1016/j.jclepro.2019.06.145

Edwards, P., & Rahman, A. (2023). *Case studies in Middle Eastern environmental policy*. Middle East Policy Council.

Farouk, S., & Al Hashimi, D. (2019). Water management in arid regions: The UAE model. *Journal of Water Resources, 58*(4), 275–289.

Ghosh, A., & Kumar, R. (2021). The task force on climate-related financial disclosures: An overview. *Climate Policy Journal, 21*(3), 409–424.

Global Reporting Initiative (GRI). (2021). *GRI standards*. https://www.globalreporting.org/standards/

Hammad, A., & Akbar, M. (2021). Environmental sustainability disclosure and firm performance: Evidence from the GCC region. *Sustainability, 13*(15), 8321. https://doi.org/10.3390/su13158321

Harrison, A., & White, L. (2022). Integrating cultural values into environmental policy. *Cultural Studies Review, 28*(2), 150–167.

Jabbour, C. J. C., Sarkis, J., Lopes de Sousa Jabbour, A. B., Renwick, D. W. S., Singh, S. K., Grebinevych, O., & Godinho Filho, M. (2020). Who is in charge? A review and a research agenda on the 'human side' of the circular economy. *Journal of Cleaner Production, 222*, 793–801. https://doi.org/10.1016/j.jclepro.2019.03.038

Khalifa, S. A., & Ahmed, M. Z. (2018). Sustainable urban planning in the Middle East. *Emirates Urban Studies, 10*(2), 112–130.

Khan, M., Serafeim, G., & Yoon, A. (2021). Corporate sustainability: First evidence on materiality. *The Accounting Review, 96*(5), 169–198. https://doi.org/10.2308/TAR-2019-0506

Liu, Y., & Thompson, J. (2019). Adapting global sustainability frameworks to local contexts. *Global Environmental Change, 29*, 123–135.

Patel, D., & Smith, J. (2022). FTSE4Good index series and corporate sustainability. *Financial Markets and Portfolio Management, 36*(1), 77–92.

Rahman, F., & Singh, G. (2020). Principles for responsible investment: A review. *Journal of Sustainable Finance, 15*(4), 450–468.

Saidi, N., & Saliba, G. (2019). *ESG and sustainable finance in the MENA region: Unlocking opportunities*. Middle East Institute. https://www.mei.edu/publications/esg-and-sustainable-finance-mena-region-unlocking-opportunities

Sarkis, J., Geng, Y., & Zhu, Q. (2020). Green supply chain management: A review and bibliometric analysis. *International Journal of Production Economics, 219*, 350–361. https://doi.org/10.1016/j.ijpe.2019.02.003

SASB. (2020). SASB standards. Sustainability Accounting Standards Board. https://www.sasb.org/standards/

Singh, H., & Patel, V. (2021). Science based targets initiative: A roadmap for corporate action. *Business Strategy and Environment, 30*(2), 321–337.

Task Force on Climate-related Financial Disclosures (TCFD). (2021). *TCFD status report 2021*. https://www.fsb.org/2021/10/2021-status-report-task-force-on-climate-related-financial-disclosures/

Turner, R., & Williams, G. (2019). Evaluating the Dow Jones sustainability indices: A comprehensive review. *Sustainability Science, 14*(5), 1319–1333.

Weber, K., & Newman, H. (2020). EU non-financial reporting directive: Implications and applications. *European Business Review, 22*(4), 408–426.

Zhang, L., & Tan, J. (2022). Case studies in global sustainability: Lessons for the Middle East. *Global Sustainability Journal, 5*(1), 55–70.

Zaabi, M., Al Marzooqi, H., & Al Hammadi, F. (2020). Environmental, social, and governance (ESG) disclosure in the UAE: An exploratory study. *Journal of Sustainable Finance & Investment, 10*(4), 321–335. https://doi.org/10.1080/20430795.2019.1705678

Chapter 13
Sustainability Maturity Assessment

Srinivasan Kada

Abstract This paper aims to address lack of uniform method to assess the maturity level of organizations on Sustainability and sustainable business practices required by local and global policies and/or law to steer the enterprise and countries toward achieving NetZero targets and limit global warming. Since COP21-Paris convention, enterprises, especially large and medium scale, are setting their NetZero target. Enterprises need to establish a baseline of their current sustainability level in order to determine science-based target and initiative to be able to achieve their goals. This demands a uniform method of assessing the maturity level irrespective of the size of the organization, the industry or the place of operations. Without this uniformity, comparison and consolidation of sustainability across sectors or countries or global comparison can be misleading due to different methods and practices employed by different companies. This paper proposes a comprehensive approach to assess maturity level of organizations across all the functions and business lines in order to establish a baseline that will provide a starting point for an organization to achieve their NetZero targets. The study combines a multi-disciplinary approach combining all functions including operations, logistics, supply chain, finance, HR, and technology and explains how to deliver the assessment in a structured way to achieve the intended objectives.

Keywords Net-zero targets · Sustainability maturity assessment · Multi-disciplinary · Sustainability development goal (SDG) · Conference of parties (COP)

Introduction

The UN now describes climate crisis as "global boiling," a term introduced by UN Chief Antonio Guterres (source: UN News, July 2023). This escalating crisis is apparent worldwide, including in the UAE, where extreme heat in September 2023

S. Kada (✉)
Du ICT—Emirates Integrated Telecommunication Company, Dubai, UAE
e-mail: srinivasan.Kada@du.ae

© The Author(s), under exclusive license to Springer Nature Switzerland AG 2026

J. Dsilva (ed.), *Redefining Sustainability*, https://doi.org/10.1007/978-3-031-88376-7_13

underscored the severity of the issue. Global temperatures often exceeded the critical 1.5 °C mark set by the Paris Agreement, indicating an earlier than expected breach of this threshold, previously projected for 2050.

In this urgent scenario, the economy and businesses are crucial. They must act decisively to align with net-zero targets, essential for preventing severe global warming by 2050. However, this journey is complex, intertwined with economic development, and without a universal solution. Businesses must evaluate their current sustainability maturity levels to establish realistic goals and investments for reaching net-zero maturity.

Understanding the initial maturity level is key to achieving a net-zero future. This assessment faces challenges such as regulatory complexities, technological barriers, supply chain issues, access to green finance, and varying environmental awareness. A comprehensive, data-driven approach is necessary, considering an enterprise's entire environmental impact. The Sustainability Maturity Assessment, grounded in the Sustainability Maturity Model, offers a framework to assist organizations in advancing toward net-zero maturity and fulfilling UN Sustainable Development Goals (SDGs).

This paper is to propose both the maturity assessment as well as a maturity model that supports an organization journey toward NetZero and attaining its UN SDG's.

The Maturity Model

The Sustainability Maturity Assessment (SMA) model is a method for evaluating how advanced a company's sustainable practices are. It serves several purposes:

1. Maps the company's current sustainability level across seven key functions.
2. Identifies reasons for unsustainable or less mature practices.
3. Sets a sustainability goal fitting the company's industry.
4. Helps pinpoint initiatives and projects to solve sustainability issues.
5. Creates a plan with milestones and timelines to reach these goals.

Developed from Sustainable SME expertise, this model addresses companies with high carbon footprints. It includes a database of over 160 questions related to company processes and roles, guiding the assessment.

Challenges

There are many challenges faced by enterprises and countries when it comes to achieving sustainability goals and targets. This is compounded by lack of standardization to assess sustainability maturity to establish current level of sustainable operations. I have cited three references to illustrate lack of standardization:

Reference 1

"The existence of multiple frameworks and methodologies with different perspectives contradicts the greater goal of global standards" Victor Chan, International Director, Global IFRS Services, Ernst & Young Global Limited.

Link: https://www.ey.com/en_id/ifrs/why-there-is-an-urgent-need-for-global-sustai nabilitystandards.

Reference 2

Top 3 challenges of Sustainability Reporting by Dimitar Grancharov, Scalefocus. In this article, the author mentions the lack of standardization and universal reporting standards. Link: https://www.scalefocus.com/blog/top-3-challenges-in-sus tainability-reporting

Reference 3

Referring to the following article by Arthur D. Little and the image clearly articulates lack of quantified sustainability goals. I attribute this challenge to lack of standardization to assess current level of maturity which is a critical input to establish targets, operations and sustainable processes, methodology and operations (Fig. 13.1).

Link: https://www.adlittle.com/en/insights/report/overcoming-challenges-sustai nability

In addition, reporting standards are varying including, to name a few, following global reporting standards (Table 13.1).

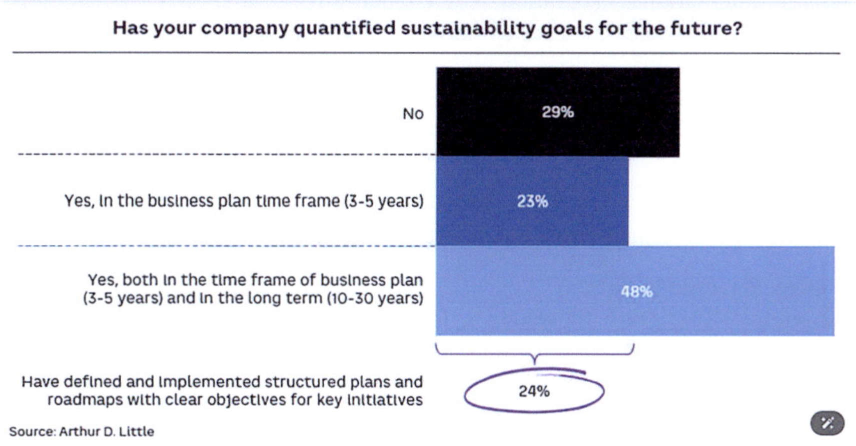

Fig. 13.1 Study from Arthur D. Lille Study on overcoming the challenges of sustainability

Table 13.1 Pros and Cons of different sustainability standards

Standards	Pros	Cons
GRI	Comprehensive coverage, flexibility, stakeholder engagement	Complexity and resource-intensive, lack of comparability
SASB	Industry-specific standards, investor focus	Limited scope, potential adoption barriers
CDP	Climate change focus, data-driven approach	Reporting burden, primarily environmental focus
ISSB	Global uniformity, integration with financial reporting	New and evolving, implementation challenges
UN Global Compact	Broad principles-based approach, global recognition	Self-reporting and compliance, lack of specificity
TCFD	Climate risk focus, decision-useful information	Narrow scope, challenges in scenario analysis
CDSB	Integration with financial reporting, consistency and comparability	Environmental emphasis, implementation complexity

Literature Review

Following research in the past highlights lack of standardization and its impact on the journey of an enterprise net zero.

IFAC Study on Sustainability Assurance

- **Source**: International Federation of Accountants (IFAC).
- **Summary**: This study reveals significant differences in sustainability assurance practices across jurisdictions. It highlights the variance in global practices for sustainability assurance, including the prevalence of assurance, level of assurance, and standard and practitioner used. This diversity points to a lack of standardization in the field.
- **Link**: *IFAC Study.*

Corporate Sustainability Maturity Model for Readiness Assessment

- **Source**: Emerald Insight.
- **Summary**: This study develops a Corporate Sustainability Maturity Model (CSMM), indicating a need for such models due to the lack of standardization. It is designed to help organizations self-assess, identify their current sustainability maturity levels, and transition into mature sustainable organizations. The research involved a systematic literature review, a pilot study, and practitioner feedback.
- **Link**: *Emerald Insight Study.*

Methodology

The Sustainability Maturity Assessment (SMA) method involves surveying key functions and roles within a company, covering areas like operations, finance, supply chain, procurement, HR, customer service, and technology.

SMA involves conducting a detailed survey of key stakeholders across all functions of an organization. Key stakeholders cover role at three levels—Executive level, Managerial level, and User level to ensure the assessment covers broader representation of people within an organization. The response to the questions is single-choice questions in order to reduce ambiguity in scoring.

This process is facilitated through a web-based application, where a company profile is set up to send invitations to key stakeholders. The company's profile determines the specific questions in the survey.

The questionnaire is created by experts (SMEs) in Enterprise Sustainability who understand its impact on business operations. After the survey, the results are compiled and compared against local, regional, and global industry benchmarks. This comparison helps a company understand its market position and plan its actions and timelines accordingly.

The assessment report aids in SWOT analysis and other evaluations like life cycle analysis (LCA) and materiality assessment, providing comprehensive insights into the company's sustainability status.

Workflow

SMA broadly follows a 10-step process to prepare the assessment report and is easier to administer this intuitive web-based application. The following picture illustrates the workflow of the assessment process (Fig. 13.2).

Fig. 13.2 10-step process to prepare the assessment report

Step	Name	Role	Description
1	Customer profile	Administrator	Set up customer profile including name, industry, key stakeholders, role, contact details
2	Select users	Administrator	Select organization user to push survey
3	Assessment page	User	Assessment questionnaire page for user to respond with appropriate selection among multiple choices
4	Consolidation page	Administrator	Consolidate all user's response for analysis and reporting
5	Analysis page	Administrator	Analyze user response for completion, trends and patterns, insights, and causes
6	Report page	Administrator	Report the analysis and pattern in visual forms like bar chart, pie chart, trend charts, spider charts, etc.
7	Heatmap	Administrator	Display heatmap of organization sustainable maturity in colors indicating maturity levels
8	Recommendation	Administrator	Provide recommendations and options to mitigate issues, address challenges and implement sustainable operations
9	Roadmap	Administrator	Provide indicative roadmap an organization can take to achieve sustainable business operations and NetZero journey
10	Actions and initiatives	Administrator	Provide indicative actions and initiatives an organization can take to improve

SMA Web-Based Application

The SMA application is a web-based application for easier administration of the assessment and has been developed as modules to provide end to end process for assessment and response.

a. Admin module—To set up, install, configure, and administer assessments.
b. Assessment module—for users to respond with their maturity level of the organization at seven core areas.
c. Analysis module—for assessment team to analyze and interpret the user responses.
d. Report module—prepare user friendly reports in the form of bar, pie, and spider charts.
e. Assessment Log module—for assessment team to track all assessment including completed, in progress and scheduled assessments

Output and Outcome

The SMA application prepares output based on the response by process stakeholders across all the functions of an enterprise and maps it to the SMA heatmap.

Before

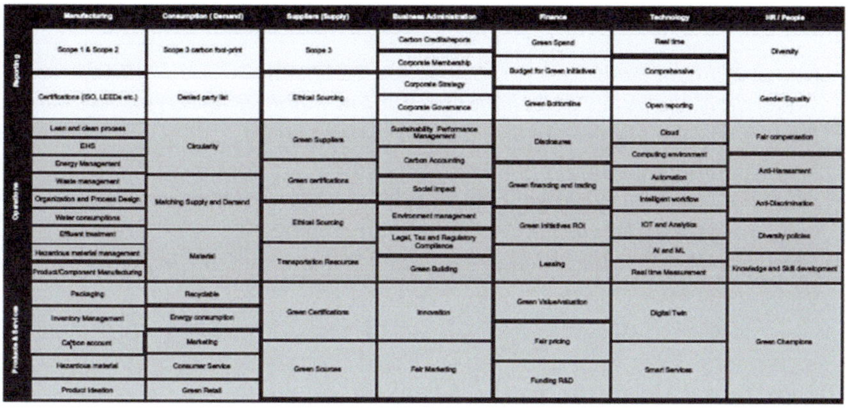

Manufacturing Consumption (Demand) Suppliers (Supply) Business Administration Finance Technology HR/People

After

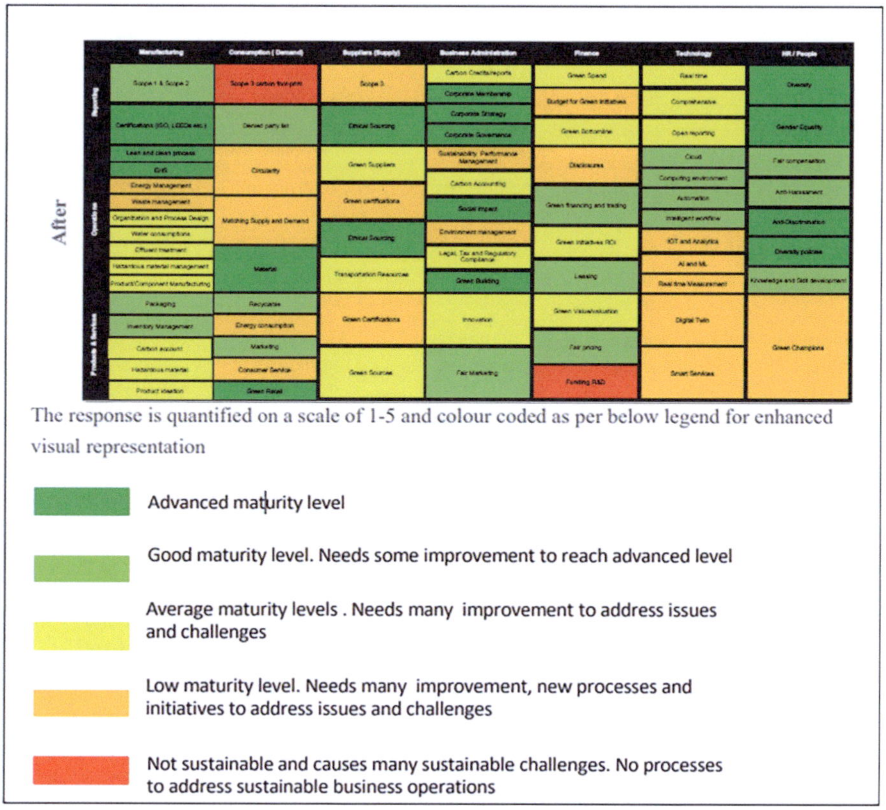

The main outcome of the assessment and heat map is to

- *Understand the maturity level by function and processes*
- *Understand the issues and challenges for low maturity levels*
- *Understand the gaps to achieve desired maturity levels*
- *Understand the causes for such maturity gaps*
- *Align the stakeholders toward sustainability maturity levels—current and target*
- *Get buy in from all stakeholders toward target maturity*
- *Identify and prioritize processes that needs maturity level improvement and/or transformation.*

Other Tools Available in the Market

The SMA has been developed as a quick assessment process to determine maturity level of the organization. The following are some of the tools available in the market. In comparison with the market tools, SMA is faster, cheaper, comprehensively covers all functions and key roles in an organization and can be conducted virtually.

Feature/Tool	Du ICT SMA	KPMG ESG maturity assessment	PwC ESG maturity diagnostic tool	EY sustainability critical assessment tool
Target audience	Cross-function and multisector	Businesses	Governments	Various sectors (industries)
Focus areas	ESG across 76 areas, 7 functions and 3 dimensions	ESG across 5 dimensions	27 ESG criteria across 5 key areas	Broad range of ESG aspects
Assessment approach	AI web-based virtual survey	Diagnostic assessment	AI-based, interactive	Part of a comprehensive service suite
Pros	AI-based for advanced analysis across all functions and stakeholders	Comprehensive ESG focus: provides rapid insights and guidance	AI-based for advanced analysis: customizable criteria	Broad industry coverage; likely comprehensive in approach
Cons	Real-time usage of tool and strengthening of question banks to be explored	May not cater to specific industry needs beyond ESG	Geared more toward government entities	Details of the tool itself are not specifically highlighted

Conclusion

In conclusion, the prevailing scenario of sustainability maturity assessment, highlighted by the significant lack of standardization as identified in key studies by the International Federation of Accountants (IFAC) and research such as the Corporate Sustainability Maturity Model (CSMM) from Emerald Insight, underscores an urgent need for a standardized approach. The proposed Sustainability Maturity Assessment (SMA) by the author is not just a requirement but a fundamental pillar in achieving a common language, process, and methodologies across organizations and countries.

The implementation of a standardized SMA is paramount for multiple reasons. First, it would ensure a uniform understanding and application of sustainability concepts, which is currently fragmented across different jurisdictions and industries. This uniformity is crucial for comparing and benchmarking sustainability efforts effectively and transparently, enabling stakeholders to make informed decisions based on reliable and comparable data.

Second, a standardized SMA would facilitate the alignment of diverse sustainability efforts with global goals, such as the United Nations Sustainable Development Goals (SDGs). By providing a common framework, it would help in harmonizing the efforts of various organizations toward these shared global objectives, enhancing the overall impact of sustainability initiatives.

Third, the adoption of a unified SMA would simplify the process for organizations, especially those operating in multiple countries, by reducing the complexity and resource burden associated with navigating through multiple, often conflicting, sustainability reporting requirements. This simplification would encourage more organizations to participate actively in sustainability reporting, increasing overall transparency and accountability in corporate sustainability practices.

Finally, the establishment of a standardized SMA is imperative in fostering a culture of continuous improvement in sustainability practices. By providing clear benchmarks and guidelines, organizations can better understand their current sustainability performance, identify areas for improvement, and track progress over time.

Therefore, the development and adoption of a standardized Sustainability Maturity Assessment as proposed by the author is not just beneficial but essential. It is a critical step toward creating a cohesive, transparent, and effective global framework for sustainability, propelling organizations, and nations toward a more sustainable, equitable, and prosperous future. This endeavor is more than a compliance exercise; it is a strategic imperative for achieving long-term sustainability goals and ensuring a resilient and sustainable world for future generations.

References

https://www.ey.com/en_id/ifrs/why-there-is-an-urgent-need-for-globalsustainability-standards, Victor Chan, International Director, Global IFRS Services, Ernst & Young Global Limited.

https://www.scalefocus.com/blog/top-3-challenges-in-sustainability-reporting, Top 3 challenges of Sustainability Reporting by Dimitar Grancharov, Scalefocus. In this article, the author mentions lack of standardization and universal reporting standards.

https://www.adlittle.com/en/insights/report/overcoming-challengessustainability, referring to the following article by Arthur D. Little and the image clearly articulates lack of quantified sustainability goals. I attribute this challenge to lack of standardization to assess current level of maturity which is a critical input to establish targets.

Chapter 14
Analyzing Sustainable Practices Impacting Project (Maintenance) Success in Power Generation in the GCC Countries

Nawab Alam

Abstract The Gulf Cooperation Council (GCC) countries have been adopting stringent measures to keep power plants running longer and better. Some of the sustainable actions adopted are making the best use of water, using nature-friendly materials, smart energy systems, and constructing green buildings. Success comes from tech improvements, money investment, rules in place, and thinking about how to be green. This investigation adopted exploratory and descriptive methods that depend on interpretivist research philosophy and inductive approach. For this purpose, interviews were carried out with 10 individuals along with thematic analysis. Findings revealed that project managers optimize resource use, minimize environmental impact, and ensure long-term viability when doing projects successfully. Another finding revealed that key practices include leveraging renewable energy sources, implementing energy-efficient technologies, optimizing water usage, and adhering to stringent waste management protocols.

Keywords Gulf Cooperation Council (GCC) · Sustainable practices · Smart energy systems · Power generation

Introduction

This study wants to look at how being careful about the environment connects with doing well on projects that keep power plants working in countries of a group called GCC. These are Bahrain, Kuwait, Oman Qatar Saudi Arabia, and United Arab Emirates. According to Aarseth et al. (2017), the important change toward making power in a way that is fair for the environment, which comes from worries about our world and laws made to protect it, matters very much here. This place is known as special when it comes to energy business (Salam & Khan, 2018). The paper

---

N. Alam (✉)
IMPZ, Dubai, UAE
e-mail: nawab.university@outlook.com

© The Author(s), under exclusive license to Springer Nature Switzerland AG 2026
J. Dsilva (ed.), *Redefining Sustainability*, https://doi.org/10.1007/978-3-031-88376-7_14

wants to investigate how fair practices make a project succeed. It aims at filling the knowledge gap by focusing only on what's different in countries from GCC region. The research has the following key objectives:

- To identify the key sustainable practices which are adopted in power generation maintenance projects within the GCC countries.
- To determine the factors which influence the successful integration of sustainable practices into power generation maintenance projects in the GCC region.
- To analyze the relationship between the adoption of sustainable practices and project success in power generation maintenance within the GCC countries.
- To assess the environmental, economic, and social impacts of sustainable practices on project success in the GCC power generation maintenance sector.
- To recommend innovative solutions and best practices for enhancing the integration of sustainable practices into power generation maintenance projects in the GCC countries, based on research findings.

Significance and Contribution

This study of the topic of "Analyzing Sustainable Practices Impacting Project (Maintenance) Success in Power Generation in the GCC Countries" is quite relevant due to its exclusive combination of three areas—sustainable practices, project maintenance, and power generation with the specific context of the Gulf Cooperation Council (GCC) countries. The research problem defined in this research is unique in the sense that it focuses on sustainable practices and their direct impact on project success, particularly in the critical field of power generation. Although most previous research studies project success and sustainability in isolation, this research aims to fill this gap by investigating the relationship between these two factors, thereby adding a comprehensive view to the discipline.

The uniqueness of the research problem is geographical context in that. The GCC countries—the rapid development countries caused by economic growth and power generation sector reliance—create a specific environment for studying sustainable practices and project success. This study attracts attention to the challenges and opportunities of a specified region considering the function of sustainable practices in risk minimization and project maintenance improving in power sector.

As far as contributions are concerned, this research is anticipated to give important perspectives on the effect of sustainable practices on project success; thus, the research fills an important gap in the existing literature. Sustainable practices that have a huge influence on project maintenance success in power generation should be identified and should then serve as a guide for industry practitioners, policymakers, and stakeholders to the GCC countries for making informed decisions and strategic planning. The study also makes contributions to academic fraternity at large since it provides the enriched knowledge on the sustainable project management practices, especially in the power generation and also provides a distinct view on the dynamics specific to the particle location of the GCC region. The importance of this research to deliver practical and context-specific recommendations and theoretical developments is in sustainable practices project success in the power generation sector in the GCC countries.

Literature Review

The review of the literature gives information about how sustainable ways and success in maintaining power generation are changing together. It highlights the big change toward making things last longer. It shows how important it is to think about money, nature, and people when a project succeeds (Elrahmani et al., 2021). It is known that in the past, people often forgot about sustainability when planning and starting projects. This lines up with new studies on business trends showing how important it is to include sustainable practices. Picking this topic is important because more interest in making energy using green practices in GCC nations (Silvius & Schipper, 2018). Factors like long-term safety for supplying power, rules, and worries about nature are driving that interest to grow. The review creates a base for learning the special problems and chances of getting sustainable power methods in GCC area.

Theoretical and Conceptual Review

This research study uses the social exchange theory recommended by Peter M. Blau, which recommends that social interactions and relationships are based upon a basic concept of reciprocity. According to this theory, people take part in social exchanges with the expectation of getting advantages that amount to or greater than the expenses they sustain. In the context of the research study, the SET can be applied to comprehend the relationships between various stakeholders associated with sustainable practices in power generation tasks in the GCC countries. The stakeholders, consisting of project supervisors, federal government authorities, and ecological companies, participate in social exchanges with the expectation of getting advantages, such as financial gains, ecological sustainability, and social acknowledgment. The theory assists in examining the characteristics of these exchanges and their influence on project success. Prior research study by Zaidan et al. (2019) has used the SET in their research studies associated with staff member engagement, highlighting its importance in comprehending social interactions in organizational contexts.

Another considerable theory utilized in this research study is the resource-based view (RBV), which was initially established by Jay B. Barney. The RBV presumes that a company's competitive advantage and efficiency are mainly driven by its special and important resources and abilities. In the context of the research study, the RBV can be applied to examine the sustainability practices embraced by power generation jobs in the GCC countries. These sustainable practices can be considered as important resources that offer a competitive advantage to the tasks. The theory enables the recognition of key resources and abilities that add to project success, such as access to renewable resource sources, effective maintenance practices, and regulative compliance.

Methodology

The research philosophy, as outlined, encompasses three primary paradigms: positivism, interpretivism, and critical realism. Interpretive way looks at what people think in a personal view by talking with them closely to understand feelings better. This way of thinking guides the decisions in gathering and checking data, affecting how research is set up as a whole. In the way things are studied, ontology and epistemology are very important parts. Ontology looks at what scientists think about existing things and truth. It asks if reality is based on personal feelings (imagining), real facts, or all of these together (practical actions). Epistemology, however, worries about how we learn and what kind of knowing is in our grasp. We separate factual knowledge you can see (positivism) from personal understanding tied to the setting around us (interpretation). These thoughts help the researcher to get a complete understanding of where and how their study is done.

The study method is explained, focusing mainly on the inductive approach. This thought process begins with a large idea, creates guesses, and checks them using real-life observation and numbers. Often it uses number methods to do so. The use of an inductive method matches the study goals to look at how green practices work in energy projects. This lets you neatly examine theories that have been given before. The study plan is explained as a helpful guide. It includes the main direction, structure, and putting into action of your research project. This way is seen as good for studying the problems of sustainable ways in power production, taking different points of view and special meanings.

In the end, the method used for looking at data is called thematic analysis. It is a popular and easy-to-use way of finding common themes or ideas in non-scientific information. This way of looking at things helps you study in detail how sustainable actions and successful projects are connected.

Findings

In the findings part, the study gives a complete report on what was found out from exploring. It shows that green actions really make power projects in GCC countries succeed. The study shows that we can reduce harm to nature, make sure things last for a long time, and use resources in the best way by following sustainable methods. The results show how important it is to encourage these actions. They help lessen waste and support sustainability in the area too. The goal of the study, which focuses on looking at how green ways impact keeping power projects going in GCC countries, is confirmed clearly. The links between eco-friendly actions and project success are looked at, focusing on the good effects for environment, money matters, and community issues.

The study shows that sustainable methods make a big difference for power maintenance projects in GCC countries. Understanding and accepting the value of ways to do things in a sustainable way is very important for doing well financially, socially, and locally. The UAE power generation business is very active in following these methods, helping to bring future gains.

The first research question asks about the main ways to be eco-friendly that are used in taking care of power generation projects within GCC countries. The results show ways to use water better, be friendly with the environment by using good materials and making smart energy systems. Also, it is important to build eco-friendly things and improve new technology for a healthier planet. The second question asks about things that can help sustain practices work well together. Important things found are—the ways technology gets better, money put in for it to grow, rules that guide action, and a sense of caring for nature forever as well as what governments decide. Checking government rules about things like technology and money is thought to be very important for winning.

The third question in the research looks at how using eco-friendly methods can help a project succeed. The study shows how important it is to use ways that help the environment when making decisions for company growth and region success. The fourth study question looks at how sustainable practices affect success in the GCC power plant maintenance area, considering environmental, economic, and social impacts.

Themes

Main themes	Description
Significance of sustainable practices	The imperative significance lies in the long-term viability of practices that prioritize environmental, social, and economic balance systems
Role of renewable sources	Renewable sources occupy a vital role in promoting sustainability
Role of government regulatory policies	Government policies play a crucial role in shaping frameworks that steer industries toward sustainable practices
Role of government regulatory framework	A well-defined regulatory framework serves as the backbone for encouraging sustainable practices
Impact of eco-friendly approaches	The adoption of eco-friendly approaches considerably influences reducing environmental footprints and mitigating resource depletion

(continued)

(continued)

Future of sustainable practices	The future holds promising prospects as sustainable practices continue to evolve

It can be summarized from the comprehensive literature review and the analysis of the results through thematic analysis that the research successfully fills the research gap regarding the lack of information on sustainable practices usage in project success in the power generation maintenance sector in GCC countries. Moreover, it can be stated that acknowledgment and realization of sustainable practice are important and play a vital role in economic, social, and regional success as it provides various benefits in the long run. According to the literature review findings and comprehensive evaluation of the respondents from the participants, it is everything the UAE power generation industry is highly proactive in adopting sustainable practices and considers it as a critical area to explore. There are these benefits or outcomes that can be acquired by the power generation industry in the UAE and the GCC countries with the adoption of sustainable practices in project success.

Conclusion

The big study of results and themes shows a different view. It is about sustainable methods in power maintenance work across GCC countries, filling the space that was not covered before. The acceptance of good ways to be green, their good effects, and important things found out helps companies in the area win for a long time.

Recommendations:

Some ideas are given to help make the power generation area more environmentally friendly. This involves putting more money into clean energy, using ways to save power better, working together on these issues in a region. Also, it is about making laws that work well and staying committed for long time with good political stability too.

Limitations of the Study

The study admits that there are restrictions, like only looking at GCC countries and using qualitative research methods. More research is needed to fix these problems. Scientists should look at other areas too and try different ways of studying things.

Future Implications

The things found out can help create policies to stop climate change, make businesses competitive, and lead GCC countries past problems with being green. The knowledge and ideas from the research are important for making future plans, rules, and approaches in keeping power plants working.

References

Aarseth, W., Ahola, T., Aaltonen, K., Økland, A., & Andersen, B. (2017). Project sustainability strategies: A systematic literature review. *International Journal of Project Management, 35*(6), 1071–1083.

Elrahmani, A., Hannun, J., Eljack, F., & Kazi, M. K. (2021). Status of renewable energy in the GCC region and future opportunities. *Current Opinion in Chemical Engineering, 31*, 100664.

Locke, J., Dsilva, J., & Zarmukhambetova, S. (2023). Decarbonization strategies in the UAE built environment: An evidence-based analysis using COP26 and COP27 recommendations. *Sustainability, 15*(15), 11603.

Salam, M. A., & Khan, S. A. (2018). Transition towards sustainable energy production—A review of the progress for solar energy in Saudi Arabia. *Energy Exploration & Exploitation, 36*(1), 3–27.

Silvius, A. G., & Schipper, R. (2018). A conceptual model for exploring the relationship between sustainability and project success. *Procedia Computer Science, 64*, 334–342.

Zaidan, E., Al-Saidi, M., & Hammad, S. H. (2019). Sustainable development in the Arab world—Is the Gulf Cooperation Council (GCC) region fit for the challenge? *Development in Practice, 29*(5), 670–681.

Chapter 15
Corporate Sustainability Practices of Publicly Listed Companies with High Sustainability Ratings in the United States

Janice Lao

Abstract Corporate sustainability has gained prominence in recent decades, yet challenges remain in its integration into business due to conflicting priorities, regulations, and guidelines. Despite research demonstrating positive links between sustainability and profits, many companies prioritize profit over society and the environment. While prior research has explored theoretical frameworks, a gap exists in a lack of concrete data and analysis on how successful companies implement key sustainability practices. This study addressed this gap by investigating key practices in the governance, strategy, and values of high-ranking sustainable firms in the United States. The results demonstrated consistent patterns across different datasets, providing valuable guidance codified in a guideline format for sustainability integration. The proposed guidelines offer a starting point to companies for assessing their position on the sustainability journey. They present a list of practices focused on efficient resource allocation, summarizing the approaches employed by high-ranking sustainable companies. The intention is not to prescribe specific actions but rather to provide guidance that will enable companies to understand and select practices based on where they are in their corporate sustainability journey.

Keywords Corporate sustainability · Governance · Boards · Sustainability leadership · Sustainability guidelines

---

J. Lao (✉)
Prescott College, Prescott, USA
e-mail: jlao@bard.edu

Bard College, New York, USA

© The Author(s), under exclusive license to Springer Nature Switzerland AG 2026

J. Dsilva (ed.), *Redefining Sustainability*, https://doi.org/10.1007/978-3-031-88376-7_15

Introduction

Over the last 20 years, corporate sustainability (CS) practices have become integral to business operations, focusing on environmental, social, and governance aspects. However, the absence of a universally accepted approach has led to confusion and tensions in balancing financial, environmental, and social objectives. The broad conceptualization of CS has made it subject to debate and ideological controversies (Bansal & DesJardine, 2014; Eccles et al., 2014). This situation is exacerbated by dynamic influences from public expectations and regulatory requirements, resulting in the proliferation of confusing guidelines and standards (Rasche et al., 2017).

CS strategies, essential for sustainable development, grapple with the alignment of economic, social, and environmental sustainability (Aragon-Correa et al., 2017; Joseph et al., 2017). Furthermore, corporate governance and board roles have become crucial in overseeing sustainability efforts (Guerrero-Villegas et al., 2018; Naciti, 2019), which has necessitated a departure from only prioritizing economic sustainability. Moreover, the persistent neoliberal emphasis on profit often sidelines societal concerns, with short-term management prioritizing immediate financial results, which hinders the integration of sustainability (Banerjee, 2008; Pal & Jenkins, 2014).

This research leveraged stakeholder theory as a foundation for its theoretical framework in recognition of the interconnectedness of corporate decisions and their diverse impacts. This aligns seamlessly with the concept of CS, which extends beyond financial metrics to encompass the well-being of a broader group of stakeholders (Gennari, 2019).

Conflicting timeframes between financial metrics and long-term environmental impacts create complexities that favor financial performance (Luo et al., 2020). Despite the lack of research challenging the idea that sustainability harms financial outcomes, current strategies prioritize shareholder value over social and environmental considerations.

While past research has highlighted the link between corporate governance and sustainability, a widening gap persists between leaders and laggards (Ioannou & Serafeim, 2019) that contributes to the confusion. Insights from sustainability leaders are crucial, which emphasizes the need for guidance within existing governance systems for effective integration.

This research drew insights from sustainability trailblazers and analyzed their approaches to transitioning toward the widespread acceptance of CS. It aimed to pinpoint the strategies employed by top-performing sustainability firms in corporate governance and strategy development. Specifically, the study answered the following research questions for U.S. companies that are highly ranked in terms of CS performance and then consolidated the findings into a set of guidelines:

1. What are the board of directors' range of practices and activities in CS-related business practices?
2. How are CS-related values enacted, as evidenced by company documents?
3. What are the CS-related dimensions within the organizational structure?

Using the content analysis methodology, this study assessed legally mandated annual reports, proxy statements, and voluntary sustainability reports—all publicly available information about the relevant U.S. companies listed throughout the 19-year history of the Corporate Knights 100 (CK100). The study identified 13 governance and strategy development indicators. The results were consistent regardless of the sector or CK100 listing duration, with the exception of the materiality matrix in sectoral composition. The findings indicated a uniform prevalence of practices among high-ranking sustainable CK100 companies, highlighting commonalities across diverse datasets.

Methodology

This study employed a summative approach to qualitative content analysis, focusing on specific words or content within the text of selected companies' public disclosures. The CK100 index served as the dataset, with a final list of 82 assessed U.S. companies. The assessment criteria were derived from the methodologies of the established four sustainability raters, namely CK100, MSCI, S&P, and Refinitiv, which ensured a comprehensive and objective evaluation. Adopting a 5-point scale introduced quantification into the content analysis, thus facilitating meaningful comparisons. Scores ranging from 1 to 5 signified the absence or presence of the desired elements. The adaptability of the scale allowed nuanced scoring for missing or inconsistent data (Hsieh & Shannon, 2005). Descriptive statistics, including frequency distribution, mode, and median, provided insights into the prevalence and distribution of sustainable practices within the dataset.

The content analysis methodology is foundational in CS research, as it offers a systematic approach to distilling intricate and opaque narratives into measurable data points (Ramos et al., 2022; Taneja et al., 2011). While its adaptability has made it valuable across various disciplines, the absence of a precise definition and standardized procedures poses challenges to its application. Qualitative content analysis involves examining textual data to reveal patterns and trends (Hsieh & Shannon, 2005). It emphasizes linguistic characteristics and contextual meaning within the text. The summative approach was employed, which involved quantifying specific words to explore their contextual usage. While offering advantages, such as unobtrusiveness, this approach may overlook broader contextual meanings.

Taneja et al. (2011) conducted a comprehensive study on CS research methodologies. Their findings revealed an empirical orientation, with 82% of the studies relying on it. The predominant use of descriptive statistics and qualitative methods highlights the method's flexibility. Studies by Ricart et al. (2005), Morgan et al. (2009), Baumann and Scherer (2010), Eccles et al. (2014), and Pal and Jenkins (2014) as well as recent research by Galbreath (2017), Henry et al. (2018), Adedayo et al. (2022), and Ramos et al. (2022) have showcased the commonality of content analysis in CS research. Utilizing publicly accessible information sources, these studies have analyzed the integration of sustainability into corporate governance and assessed the implementation of sustainability guidelines.

This study is crucial for advancing the mainstream adoption of CS as it examined key areas of corporate governance and strategy development in companies that rank highly in terms of sustainability. Through addressing gaps highlighted by existing literature, this study enriches theory by exploring how these companies implement sustainability practices, thereby offering practical insights. The choice of content analysis was apt for extracting tangible insights from documents and rigorously analyzing textual information, which enhanced this study's contribution to the understanding of sustainability's integration into corporate practices (Ramos et al., 2022). The alignment with theoretical gaps and the methodological appropriateness of content analysis positions this study as a valuable contribution to CS knowledge.

Findings

This study identified 13 indicators executed primarily on governance processes and in the strategy development stage. Regardless of (1) sectoral composition or (2) the length or frequency of listing on the CK100, except for the materiality matrix indicator for sectoral composition, the pattern of results was similar to the most prevalent practice within the research dataset. These results suggested that CK100 high-ranking sustainable companies have clear prevalent practices that were similar across the different datasets. This was confirmed through a comparison of the median and mode results, which revealed an equal distribution for most of the indicators, except for these indicators: materiality matrix, the others and non-climate change commitment, and the chief sustainability officer (CSO), which exhibited some inconsistent distribution results. Table 15.1 summarizes and outlines the key findings in a guideline format along with high-level quantitative results.

Conclusion

Through an examination of three aspects of corporate governance and strategy development, this study found that prevalent practices exist for nearly all 13 of the identified indicators. Moreover, achieving frequent high rankings as a sustainable company in a particular sector may not be a prerequisite for implementing these practices. Therefore, the path to becoming a sustainable business is attainable today through the adoption of the proposed guidelines as a starting point. This research is significant as it addressed a research gap in the form of a lack of existing data and analysis on how these critical areas are implemented in practice. This research assessed actual data on successful implementation practices.

This research offers comprehensive guidelines derived from an analysis of high sustainability companies, which serve as a valuable resource for companies at any

Table 15.1 Summary of prevalent practices on corporate sustainability from high-ranking sustainable US-listed companies

Indicators	Most prevalent practice	Frequency	Additional context
Q1: Role of the board of directors in sustainability-related business practices			
Board committees	Board committee not specific to corporate sustainability (CS), without a publicly available board charter	49%	Nearly 25% of companies had a board committee specific to CS and with a publicly available board charter
Named Executive Officer (NEO) compensation	No NEO compensation linked to CS	Majority	Nearly 40% of companies had this practice in place
Chair and chief Executive Officer (CEO)	No separation but with lead independent director	Majority	Aligned with New York Stock Exchange (NYSE) and Nasdaq listing requirements
Board independence	Majority comprising independent directors	Majority	Aligned with NYSE and Nasdaq listing requirements
Board diversity, equity and inclusion (DEI)	Comprising at least 30% female independent directors and at least one director with racial or sexual orientation diversity	Majority	Nearly 40% of companies had at least one female director, or at least one director with racial or sexual orientation diversity, regardless of independence status
Q2:CS-related values and culture			
CS policies	Overarching and comprehensive CS policy	40%	Over 50% of companies did not have an overarching and comprehensive CS policy but instead had at least one or two CS related policies
Reporting	Yearly reporting as part of an integrated sustainability report and discussion in either the annual report or proxy statement	Majority	All companies in the data set publicly reported their CS performance
Sustainability definition	Companies in the data set defined CS similarly to the triple bottom line concept	Majority	9% did not define CS, while 4% defined it similarly to the Brundtland Commission's definition, i.e. temporal nature with trade-offs

(continued)

Table 15.1 (continued)

Indicators	Most prevalent practice	Frequency	Additional context
Materiality matrix	Matrix-style aligned with third-party standard with description	34%	Approximately 80% of companies provides some disclosures on their material or priority topics; only approximately 48% of companies presented it in a matrix, presenting not only it in a matrix presenting not only business impact but also external stakeholder views
Climate change (CC) commitment	Credible third party-approved long-term CC targets	Majority	35% had long-term CC targets but they were not submitted to or approved by a third party
Others & non-CC commitment	At least two other non-CC commitments that refer to or align with third-party standards or guidelines	43%	26% had at least one other non-CC commitment that referred to or aligned with third-party standards or guidelines
Q3:CS-related dimensions within the organizational structure			
Management committee and sustainability team	Presence of both management committee and sustainability teams	Majority	13% did not have either
CSO	Presence of a CSO from a nonmarketing field reporting directly to the NEO or board	Majority	31% did not have a CSO at all

stage of their CS journey. The research's significance lies in its provision of actionable insights for various stakeholders, including boards of directors, shareholders, senior management, CS professionals, and business students. The guidelines emphasize the efficient allocation of financial resources by summarizing practices implemented by successful, sustainable companies. Encouraging companies to shift their focus from mere financial gains, the guidelines promote a heightened emphasis on the delivery of environmental and social benefits to all stakeholders. This data-driven guidance enhances the understanding of the organizational landscape and its requirements, thus supporting the mainstreaming and expansion of CS initiatives. By drawing insights from successful models, this research serves as a starting point for companies that seek to adopt sustainable practices, fosters a holistic integration that goes beyond performative actions, and prioritizes authentic and impactful change.

This study has some limitations, such as its exclusive focus on publicly held large U.S. corporations, omitting nonprofits. Methodologically, the study relied on accessible public disclosures, which hindered a comprehensive analysis of non-listed companies. Data extracted from public documents pose limitations in the assessment of CS, potentially leading to greenwashing. Furthermore, this study's focus on governance and strategy development excluded strategy implementation and the impact of financial performance. Moreover, a lack of universally accepted definitions of sustainability and a reliance on nominal data and descriptive statistics present further constraints. Despite these limitations, this study provides insights into sustainable corporate governance trends and thus serves as a foundation for future research.

References

Adedayo, E. O., Adex, B. O., & Oyewo, B. (2022). Sustainable development goals (SDG) reporting: An analysis of disclosure. *Journal of Accounting in Emerging Economies, 12*(5), 761–789. https://doi.org/10.1108/JAEE-02-2020-0037

Aragon-Correa, J. A., Marcus, A., Rivera, J., & Kenworthy, A. (2017). Sustainability management teaching resources and the challenge of balancing planet, people, and profits. *Academy of Management Learning & Education, 16*(3), 469–483. https://doi.org/10.5465/amle.2017.0180

Banerjee, S. B. (2008). Corporate social responsibility: The good, the bad and the ugly. *Critical Sociology, 34*(1), 51–79. https://doi.org/10.1177/0896920507084623

Bansal, P., & DesJardine, M. R. (2014). Business sustainability: It is about time. *Strategic Organization, 12*(1), 70–78. https://doi.org/10.1177/1476127013520265

Baumann-Pauly, D., & Scherer, A. G. (2010). *MNEs and the UN Global Compact: An empirical analysis of the organizational implementation of corporate citizenship.* IOU Working Paper No. 114. https://doi.org/10.2139/ssrn.1705532

Eccles, R., Ioannou, I., & Serafiem, G. (2014). The impact of corporate sustainability on organizational processes and performance. *Management Science, 60*(11), 2835–2857. https://doi.org/10.1287/mnsc.2014.1984

Galbreath, J. (2017). The impact of board structure on corporate social responsibility: A temporal view. *Business Strategy and the Environment, 26*(3), 358–370. https://doi.org/10.1002/bse.1922

Gennari, F. (2019). How to lead the board of directors to a sustainable development of business with the CSR committees. *Sustainability: Science Practice and Policy, 11*(24), 69–87. https://www.mdpi.com/2071-1050/11/24/6987

Guerrero-Villegas, J., Pérez-Calero, L., Hurtado-González, J., & Giráldez-Puig, P. (2018). Board attributes and corporate social responsibility disclosure: A meta-analysis. *Sustainability, 10*(12), 1–22. https://doi.org/10.3390/su10124808

Henry, L. A., Buyl, T., & Jansen, R. J. G. (2019). Leading corporate sustainability: The role of top management team composition for triple bottom line performance. *Business Strategy and the Environment, 28*(1), 173–184. https://doi.org/10.1002/bse.2247

Hsieh, H.-F., & Shannon, S. E. (2005). Three approaches to qualitative content analysis. *Qualitative Health Research, 15*(9), 1277–1288. https://doi.org/10.1177/1049732305276687

Ioannou, I., & Serafeim, G. (2019). Corporate sustainability: A strategy? *Harvard Business School Accounting & Management Unit Working Paper, 19*(65). https://doi.org/10.2139/ssrn.3312191

Joseph, J., Borland, H., Orlitzky, M., & Lindgreen, A. (2017). Seeing versus doing: How businesses manage tensions in pursuit of sustainability. *Journal of Business Ethics, 164*(1), 349–370. https://doi.org/10.1007/s10551-018-4065-1

Klettner, A., Clarke, T., & Boersma, M. (2014). The governance of corporate sustainability: Empirical insights into the development, leadership and implementation of responsible business strategy. *Journal of Business Ethics 122*(1), 146–165. https://doi.org/10.1007/s10551-013-1750-y

Luo, B., Tang, Y., Chen, E., Li, S., & Luo, D. (2020). Corporate sustainability paradox management: A systematic review and future agenda. *Frontiers in Psychology, 11*, 1–15. https://doi.org/10.3389/fpsyg.2020.579272

Morgan, G., Ryu, K., & Mirvis, P. (2009). Leading corporate citizenship: Governance, structure, systems. *Corporate Governance, 9*(1), 39–49. https://doi.org/10.1108/14720700910936047

Naciti, V. (2019). Corporate governance and board of directors: The effect of a board composition on firm sustainability performance. *Journal of Cleaner Production, 237*, 1–7. https://doi.org/10.1016/j.jclepro.2019.117727

Pal, M., & Jenkins, J.J. (2014). Reimagining sustainability: An interrogation of the Corporate Knights' Global 100. *Environmental Communication, 8*(3), 388–405. https://doi.org/10.1080/17524032.2014.906477

Ramos, D., Chen, S., Rabeeu, A., & Rahim, A. (2022). Does SDG coverage influence firm performance? *Sustainability, 2022*(14), 1–10. https://doi.org/10.3390/su14094870

Rasche, A., Morsing, M., & Moon, J. (Eds.). (2017). *Corporate social responsibility: Strategy, communication*. Cambridge University Press.

Ricart, J., Miguel, Á. R., & Sánchez, P. (2005). Sustainability in the boardroom: An empirical examination of Dow Jones Sustainability World Index leaders. *Corporate Governance: The International Journal of Business in Society, 5*(3), 24–41. https://doi.org/10.1108/14720700510604670/full/html

Taneja, S. S., Taneja, P. K., & Gupta, R. K. (2011). Researches in corporate social responsibility: A review of shifting focus, paradigms, and methodologies. *Journal of Business Ethics, 101*(3), 343–364. https://doi.org/10.1007/s10551-010-0732-6

Chapter 16
Zero Waste Goals and TRUE Challenges in the UAE

Elena Babii

Abstract This white paper addresses the persistent global challenges of waste management, focusing on the shift toward Zero Waste practices and the adoption of TRUE (Total Resource Use and Efficiency) certification in the United Arab Emirates (UAE) and globally. This document underscores the urgency of a global transition by emphasizing environmental impact, resource conservation, and waste crisis. As nations urbanize and economies grow, waste generation is projected to rise significantly, requiring a strategic approach to mitigate environmental degradation. The objective of this paper is to explore municipality solid waste (MSW), highlighting global contributions and statistics, while analyzing the UAE's commitments, efforts, and challenges in waste management. This paper emphasizes the importance of the TRUE certification, which serves as a guideline for corporations in the UAE, providing a framework for measuring, improving, and certifying waste management performance. This paper recognizes the adoption of the TRUE certification across various sectors in the UAE, aligning with international sustainability standards. The methodology used in this research involves a combination of qualitative and quantitative methods. The key findings of this research identified and categorized TRUE challenges into technical, economic, policy, regulatory, and cultural aspects, followed by recommendations. While achieving absolute Zero Waste is a considerable challenge, successes have been demonstrated globally. In the UAE, a historic moment took place at COP28, where local authorities committed to decarbonize waste management and establish a circular economy platform, further solidifying the UAE's commitment to sustainability. This paper encourages collaboration and innovation to accelerate progress and achieve sustainable waste management globally.

Keywords Zero Waste · Waste management · TRUE certification · Sustainability · Circular economy

---

E. Babii (✉)
AESG, Dubai, UAE
e-mail: e.babii@aesg.com; babiielena91@gmail.com

© The Author(s), under exclusive license to Springer Nature Switzerland AG 2026
J. Dsilva (ed.), *Redefining Sustainability*, https://doi.org/10.1007/978-3-031-88376-7_16

Introduction

Scope and Objectives

The aim of this paper is to address the challenges and solutions of MSW, focusing particularly on the adoption of Zero Waste practices and TRUE guidelines in the UAE and worldwide. This paper analyzes key waste management statistics globally but with a specific focus on the UAE. The objectives are (1) to highlight the urgency of a global transition toward sustainable waste management practices due to environmental impact, (2) to analyze the UAE's commitment, efforts, and challenges in a rapidly developing urban environment with responsible waste management practices, (3) to highlight the significance of TRUE certification as a guideline for measuring waste management challenges in the UAE, and (4) to recommend solutions to overcome the identified challenges, including policy support, data collection, and reporting transparency.

The Importance of Waste Management Globally and in the UAE

The global conversation surrounding waste management has undergone a transformative shift, with a call for the adoption of Zero Waste practices being discussed at governmental levels. Championed by various authoritative bodies, such as the United Nations Environment Programme (UNEP),[1] the World Bank Group (WBG),[2] and the Ellen MacArthur Foundation (EMF),[3] Zero Waste has emerged as a critical initiative in mitigating environmental degradation, conserving resources, and changing the alarming increase of waste worldwide.

As stated in the World Bank report "What a Waste 2.0: A Global Snapshot of Solid Waste Management to 2050":

> By 2050, the world is expected to generate 3.40 billion tons of waste annually, increasing drastically from today's 2.01 billion tons. (Kaza et al., 2018)

In this section, we will explore the global waste dynamics, emphasizing regional contributions and key statistics that support the challenges and opportunities in contemporary waste management. From high-income countries with disproportionate waste generation to the distinct patterns in specific regions. Statista[4] provides

---

[1] The global authority for the environment with programs focusing on climate, nature, pollution, sustainable development, and more.

[2] With 189 member countries, the World Bank Group is a unique global partnership fighting poverty worldwide through sustainable solutions.

[3] A UK registered charity which promotes the circular economy.

[4] The Statistics Portal for Market Data.

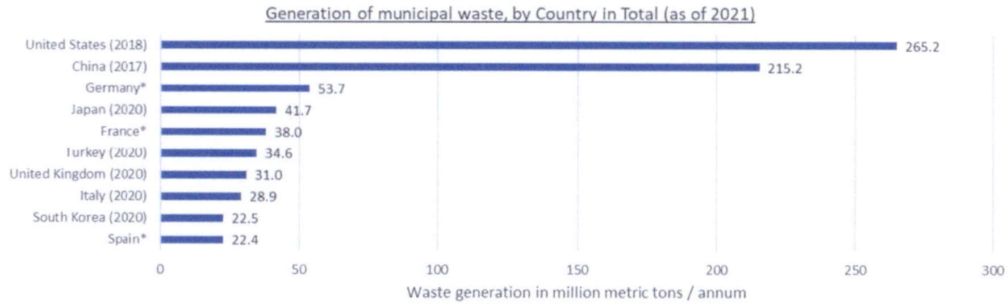

Fig. 16.1 Statista, Generation of MSW worldwide as of 2021, by select country

valuable insights on the total municipal waste generated and the waste generated per capita. This information and data should inform strategic initiatives to be taken for a more sustainable and responsible approach to waste management on a global scale.

Largest Waste Producers Worldwide

There are two main ways to look at waste production: total amount and per capita. Each analysis reveals different aspects of the issue (Fig. 16.1).

Largest Waste Producers of Municipal Solid Waste (MSW),[5] by Total Amount Worldwide

The below chart informed by Statista represents that:

> The United States and China are the largest producers of municipal waste in the world, generating over 200 million metric tons, as of the latest data available. In contrast, municipal waste generation in the UK stood at around 31 million tons in 2020 (Alves, 2023a, 2023b, 2023c).

Largest Waste Producers of MSW Per Capita Worldwide

The below chart informed by Statista represents that: "In 2021, per capita municipal waste generation in Norway was estimated at nearly 800 kg. This made the Nordic country one of the largest per capita municipal waste producers worldwide. Most

---

[5] The composition of municipal solid waste varies greatly from municipality to municipality, and it changes significantly with time. Most definitions of municipal solid waste do not include industrial wastes, agricultural wastes, medical waste, radioactive waste, or sewage sludge.

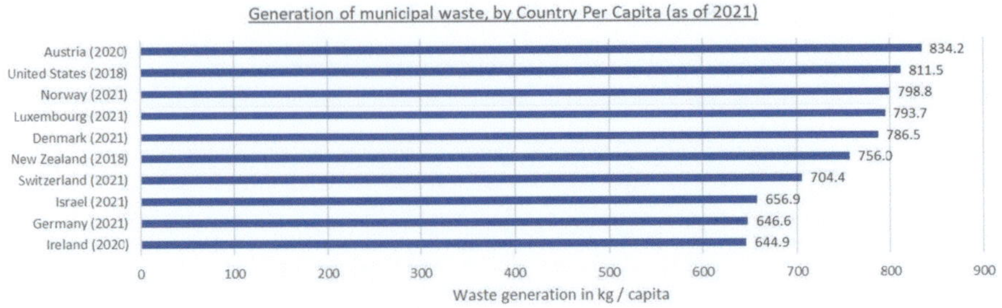

Fig. 16.2 Statista, per capita generation of municipal waste worldwide as of 2021, by select country

recently in 2018, municipal waste generation in the United States averaged 811.5 kg per capita" (Alves, 2023a, 2023b, 2023c) (Fig. 16.2).

The core observation from 2018 to present is that high-income countries generate more waste on both, a total amount produced and on a per capita basis. By the numbers, high-income countries generate approximately 34% of global MSW but only represent approximately 16% of the global population (Kaza et al., 2018).

It is worth mentioning that high-income countries, like some of those in the graphs above, have well-established waste management facilities and strategies. Although this subject is not covered in this paper, it leaves room for further necessary research.

Middle East Region

Middle East: The region produces approximately 150 million tons of municipal solid waste (MSW) annually, with Saudi Arabia and the UAE being the largest contributors. Individuals in the Middle East generate more waste than the global average (Maheshwari, 2022).

Focus on the United Arab Emirates (UAE)

The UAE exceeds 21 million tons of MSW annually. The Ministry of Climate Change and Environment (MOCCAE) in the UAE states that per capita municipal solid waste generation reached about 2.1 kg/day before falling to around 1.8 kg/day, as a result of the enormous efforts made by the relevant authorities in the UAE. Converting this to kilograms per year (assuming 365 days) gives us a range of 657–667 kg/year; as a result, this would position the UAE in the list of the highest MSW producers per capita globally.

While not explicitly listed in the above graphs, the UAE's per capita waste generation is undoubtedly significant and comparable to some of the worst offender countries measured. Depending on the specific study, methodology, year, and or report used to compile the top list, the data for the UAE might not have been readily available or included in the analyses at that time. In addition, some reports might target a specific regional focus or choose countries with well-established and consistent data records and exclude rapidly developing economies.

Overall, the UAE's waste generation per capita is comparable to other high-income countries and requires attention and urgent action for sustainable waste management practices.

Theoretical Underpinning

Overview

The theoretical underpinning of this research paper revolves around the concept of Zero Waste and its significance in the context of global waste management practices.

The urgency of transitioning toward sustainable waste management practices worldwide is emphasized through the analysis of key waste management statistics, globally and regionally with a particular focus on the UAE. Theoretical underpinning of this paper draws upon established best practices in sustainability, environment, and resource conservation to underscore the need for a global shift toward responsible waste management. Furthermore, the analysis of the UAE's commitment, efforts, and challenges in waste management draws upon theories of policy implementation and urbanization, exploring how government policies and rapid urban development influence waste generation and disposal. Additionally, the Zero Waste principles and TRUE (Total Resource Use and Efficiency) rating system served as a framework for analyzing waste management challenges and proposing solutions. TRUE certification, administered by the Green Business Certification Inc. (GBCI), provides businesses with a comprehensive approach to measure, improve, and certify their waste management performance, aligning with international sustainability standards. By contextualizing the urgency of transitioning toward sustainable waste management practices within theoretical frameworks of Zero Waste, Policy Implementation and TRUE certification, this research paper aims to contribute to the discourse on effective waste management strategies globally and in the UAE.

I explore and elaborate in detail on the theoretical underpinning of this research paper in the following sub-sections.

What is Zero Waste?

The definition of Zero Waste according to the Zero Waste International Alliance (ZWIA)[6] states:

> Zero Waste: The conservation of all resources by means of responsible production, consumption, reuse, and recovery of products, packaging, and materials without burning and with no discharges to land, water, or air that threaten the environment or human health (ZIWA, 2022).

The Zero Waste concept entails a holistic approach, emphasizing multiple behavioral changes for the end user that encompass stages, such as repair, redesign, rethink, refuse, repurpose, re-gift, resiliency, responsibility, relationship, and the well-known RRR principles of reduce, reuse, and recycle.

Why do we need Zero Waste? The urgency of embracing Zero Waste comes from the escalating environmental crises, including climate change, pollution, and resource depletion. These are all discussed and highlighted by international authorities, such as the United Nations (UN), the World Health Organization (WHO), and environmental agencies.

Initiatives on Waste Management and Key Stakeholders in the Waste Management Industry in the UAE

Governmental Initiatives

Among this important global movement, the UAE has eagerly embraced sustainable initiatives, seeking to align the country's trajectory and execution with global sustainability best practices.

Local authorities in the UAE, such as the Dubai Municipality and Tadweer,[7] have taken proactive measures to promote sustainable practices as a testament to the UAE's commitment to a more sustainable future.

In 2019, Abu Dhabi Waste Management Centre, Tadweer, signed a Memorandum of Understanding with the Green Business Certification Inc. (GBCI),[8] to collaborate in the field of Zero Waste management. This Memorandum will support Tadweer's efforts to minimize the waste production of the UAE by pursuing Zero Waste goals as a strategy for full diversion of waste from non-recovery solutions, while creating a responsible and sustainable society that adopts and implements Zero Waste lifestyle choices.

---

[6] An alliance working toward a world without waste through public education and practical application of Zero Waste.

[7] Abu Dhabi Waste Management Centre.

[8] An American organization that provides third-party credentialing and verification for several rating systems relating to the built environment.

Fig. 16.3 Source—UAE circular economy policy

In 2021, the UAE took significant steps toward adopting a circular economy model, recognizing its potential for economic growth, environmental sustainability, and resource security. The "UAE Circular Economy Policy"[9] outlines a strategic framework to implement the circular economy across various sectors. It prioritizes infrastructure, sustainable manufacturing, sustainable transportation, sustainable food production, and consumption and promotes innovation and collaboration.

The image below illustrates the primary sectors where circular economy principles are implemented, delineating the four priority sectors, and identifying eight overarching priorities (Fig. 16.3).

"The UAE government is committed to achieving more Sustainable Consumption and Production (SCP) and meeting SDG 12 by moving away from a degenerative linear (take, make, use, dispose) economy" (Government of Dubai, 2021). This commitment to SCP and the Circular Economy plays an important role in the UAE's Vision and the UAE's Green Agenda 2030.

COP28

In addition to that, a historic moment took place on the 6th of December 2023 in Dubai at COP28[10] where the MOCCAE and Tadweer officially launched the "Waste to Zero"[11] initiative. This global endeavor, unveiled during COP28, aims to champion

[9] The policy is a framework for identifying the priorities in terms of circular economy. Priorities include infrastructure, sustainable transportation, sustainable manufacturing, sustainable food production, and consumption.

[10] The 2023 United Nations Climate Change Conference or Conference of the Parties of the UNFCCC is more commonly known as COP28.

[11] The "Waste to Zero" initiative will host a series of discussions, workshops, and awareness programs worldwide, focusing on sustainable resource management, circular economy objectives, and innovative waste management and carbon emission reduction technologies.

efforts in decarbonizing Waste Management and establishing a circular economy platform, further solidifying the UAE's commitment to sustainability. The UAE government has also launched several initiatives supporting Zero Waste goals, including the Waste to Wealth Strategy.[12] These initiatives create a supportive environment for companies pursuing Zero Waste ambitions.

Key Stakeholders

The ambitious pursuit of Zero Waste in this rapidly growing region comes with its own unique set of challenges, particularly in supporting population growth while adapting and improving environmentally responsible waste management practices. The encountered challenges vary from infrastructure limitations and cultural shifts in consumer behavior to regulatory frameworks that demand innovative solutions and collaborative actions.

Recognizing the need for a structured approach to Zero Waste, many corporations in the UAE are turning to third-party certifications like Total Resource Use and Efficiency. This certification, administered by the GBCI, provides a comprehensive framework for businesses to measure, improve, and certify their waste management performance. Corporates view TRUE certification as a strategic tool not only to align with international sustainability standards but also to demonstrate transparency and accountability to their stakeholders.

TRUE concepts have been adopted across various sectors including but not limited to hospitality, manufacturing, commercial offices, retail centers, and data centers in the UAE. Therefore, TRUE certification is seen as a tangible demonstration of their dedication to sustainable practices.

Corporate Adherence to Zero Waste Movement and TRUE Certification

Not long ago, the majority of people paid little attention to waste; it was often overlooked and was rather automatically referred to as an "out of sight, out of mind" paradigm. As long as trash bins were emptied, people did not care to question or think about where the waste ended up. However, this mindset has changed. There is now a global awareness that the handling of waste is crucial, and corporates are becoming more mindful of their waste generated and its final destination.

Over the past decade, numerous prominent corporations and brands have significantly advanced their sustainability efforts by aligning with the Zero Waste movement. These strides involve eliminating manufacturing waste sent to landfills, with

[12] An ambitious national program aiming to divert 75% of waste from landfills by 2025 through advanced recycling, composting, and waste-to-energy projects.

more than 90% diverted from both landfills and incinerators. This success is most notable among companies responsible for the manufacturing of well-known household items.

More recently, enterprises leading the Zero Waste movement are ranging from small businesses to large corporations and are actively engaged in efforts, such as diverting manufacturing waste from landfills and empowering consumers with tools to reduce household waste. A few examples of the enterprises advocating Zero Waste are Unilever, Procter & Gamble, Google, Subaru, Mars, etc. Together, these economic titans are leading the way toward achieving a Zero Waste future. As per USGBC,[13] organizations committed to closing the loop on waste through Zero Waste ambitions are moving beyond single facilities and committing to Zero Waste targets across their portfolios of assets by using TRUE certification to advance their net zero goals.

> Leading the charge are companies like Colgate-Palmolive and Mastercard. To meet their goals, these household names have turned to a new program by TRUE, the comprehensive Zero Waste certification program led by GBCI (Plockova, 2023).

Trends are changing, and waste is not welcome anymore. Companies from many industries are diving headfirst into TRUE certification, proving or at least making strong efforts to demonstrate that net Zero Waste might soon be the new way of doing business.

Who is Pursuing TRUE Zero Waste in the UAE?

Several companies are actively pursuing Zero Waste practices and implementing strategies associated with the TRUE certification criteria. Some examples include Unilever (MENA) which implements eco design principles, reduces packaging waste, and participates in recycling programs to minimize their environmental footprint; Majid Al Futtaim which launched a comprehensive sustainability strategy focused on reducing waste across their diverse businesses, including shopping malls, cinemas, and entertainment centers; Emirates Airlines which is working to reduce inflight waste through sustainable packaging, food waste reduction programs, and exploring biofuel alternatives; and DP World (Jebel Ali Port) which is investing in sustainable technologies and infrastructure to minimize waste generation and pollution within their ports and logistics operations.

Several waste management companies like Tadweer (Abu Dhabi), Bee'ah (Sharjah), and Dulsco (Dubai) are leading the charge by showcasing significant efforts in waste reduction, recycling, and resource recovery, putting them on a potential path toward

Zero Waste practices which are derived from the TRUE standard.

While several companies are actively pursuing waste reduction and resource recovery which are aligned with Zero Waste principles, achieving and maintaining

---

[13] Is a nonprofit organization that supports the development of prosperous, healthy, and resilient communities through the transformation of the built environment?

this demanding TRUE certification presents several challenges, which we explore in the next section of this paper, followed by potential solutions.

It is important to note that as TRUE certification is still relatively new and evolving, there is not a publicly available or a comprehensive list of companies in the UAE pursuing this certification (yet).

Similarly, no information is currently available on companies in the UAE that have obtained the TRUE certification. Achieving TRUE certification can take time, often requiring a phased approach with initial pre-certification followed by full certification.

Zero Waste Principles and Best Practices

Zero Waste best practices encompass the principles of reduce, reuse, recycle, recovery, and disposal, aiming to minimize waste generation and environmental impact. By prioritizing waste prevention, material reuse, efficient recycling processes, resource recovery, and responsible disposal methods, Zero Waste initiatives strive to create a sustainable and circular economy (Fig. 16.4).

In addition to the core principles of Zero Waste (reduce, reuse, recycle, recovery, and disposal), TRUE certification incorporates additional practices to further promote sustainability and waste reduction. These practices include:

1. Source Separation: Implementing effective systems to separate different types of waste at the source, facilitating easier recycling and diversion from landfills.
2. Composting: Encouraging the composting of organic waste materials to divert them from landfills and create valuable soil amendments.
3. Material Reclamation: Prioritizing the reclamation and reuse of materials from discarded items, promoting a circular economy, and reducing the need for raw resources.

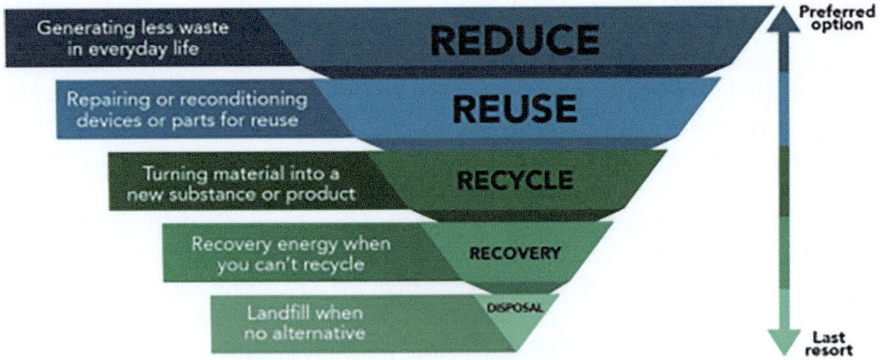

Fig. 16.4 Responsible waste management hierarchy by SG

4. Sustainable Purchasing: Adopting procurement practices that prioritize environmentally friendly and reusable products, reducing waste generation from the outset.
5. Education and Training: Providing education and training programs to employees and stakeholders to raise awareness about waste reduction strategies and promote participation in waste management initiatives.

By incorporating these additional best practices, TRUE certification goes beyond Zero Waste principles to promote comprehensive resource efficiency and environmental stewardship within organizations and communities.

Methodology

This research paper employs a mixed-methods approach, combining qualitative and quantitative research methods to comprehensively investigate the challenges associated with achieving TRUE certification in the UAE. Some of the challenges associated with achieving TRUE certification in the UAE have been meticulously examined and assessed through (1) in-person assessments of the projects site. This qualitative method involves visiting the project site to observe and evaluate various aspects, such as operations and environmental conditions. This method allows us to gather firsthand information, and impressions that may not be captured through other data collection methods. (2) The quantitative method was applied during waste auditing. This quantitative method provides measurable data on the types and amounts of waste generated, which is essential for assessing current waste management practices and identifying areas for improvement. (3) A gap analysis was carried out which involved comparing current waste management practices at the project site against established sustainability standards, in our case TRUE rating systems and Zero Waste best practices. The gap analysis helps identify discrepancies between current practices and desired goals, which informs the development of strategies and recommendations for improvement. (4) A review of existing policies was undertaken. This method involves reviewing and analyzing existing policies, regulations, and standards related to waste management at the project site.

Below is a detailed overview of each method utilized:

Quantitative Approach

Quantitative data was collected through waste audits and surveys conducted at the project site. The waste audits involved detailed worksheets provided by the waste management vendor, containing measurable data on various waste streams, such as general waste, cardboard, aluminum, mixed plastic, and paper. These audits were conducted over a period of 4 consecutive months, with a sample size of 5 waste streams. The collected data was precisely reported to both the client (remains anonymous) and the Sustainability Consultancy firm for further analysis. This approach facilitated a quantitative assessment of current waste management practices, offering insights into the types and quantities of waste generated on site.

Qualitative Approach

Qualitative data collection involved in-person assessments of the project site conducted by the Sustainability consultant. These assessments allowed for the direct observation and evaluation of operational and environmental conditions at the site. Additionally, semi-structured interviews and meetings were conducted with participants from various departments including operations, procurement, human resources, facility management, and building tenants. The qualitative component aimed to gain deeper insights into participants' experiences, perceptions, and attitudes toward waste management practices within the corporate and the broader context of waste management in the UAE. To ensure diversity in perceptions, a sampling sheet in the form of a sustainability matrix was developed for each discipline. Due to the confidentiality of the project, sampling sheets are available on request. Interviews were conducted both in-person and virtually to accommodate participants' preferences and ensure accessibility.

Participant Profile

The participants involved in this study encompassed a range of stakeholders engaged in waste management practices in the UAE, including project managers, environmental consultants, industry experts, and representatives from various corporate departments. The selection criteria were based on participants' expertise and involvement in waste management projects relevant to the research objectives, ensuring a comprehensive understanding of the challenges and opportunities in achieving TRUE certification.

Research Approach

The research approach employed in this study leans toward a deductive approach, starting with the formulation of hypotheses based on existing theories and empirical evidence related to waste management practices and sustainability standards. However, an inductive approach was also incorporated during data analysis to explore emerging themes and patterns derived from qualitative data.

Research Philosophy

This study aligns with an interpretivist research philosophy, emphasizing the importance of understanding the subjective meanings and interpretations to their experiences. By adopting an interpretivist stance, this research seeks to explore the complexities of waste management practices in the UAE from diverse perspectives.

Sampling Technique

The sampling technique utilized in this study was goal-directed sampling, whereby participants were purposefully selected based on their relevance and expertise in the field of waste management in the UAE. This approach ensured that the data collected were both relevant and informative for addressing the research objectives.

Overall, the combination of in-person assessments, waste audits, gap analysis, and policy reviews allow researchers to gather comprehensive data from multiple sources and perspectives. This mixed-method approach enables a thorough analysis of waste management practices, challenges, and opportunities, as well as the formulation of evidence-based recommendations for improvement.

Findings

Zero Waste and TRUE Challenges and Recommendations

Technical

- Limited Infrastructure: While there is rapidly developing waste management infrastructure in the UAE, there still lacks sufficient facilities for various advanced recovery and processing technologies required for Zero Waste and TRUE certification. A few examples of infrastructure investments that are still necessary to be introduced to this market include large-scale organic waste composting facilities and specialized plastic recycling plants.
- Technology Adoption: Implementing innovative waste management technologies (e.g., anaerobic digestion, AI-powered sorting systems, etc.) requires significant capital investment, overcoming initial difficulties in operational integration and expertise.
- Data Collection and Monitoring: Accurately measuring and monitoring waste streams throughout the entire lifecycle, from generation to disposal, requiring robust data management systems and skilled personnel for implementation.

Economic and Business

- Costs and Investments: Comprehensive recycling and composting often have higher upfront costs compared to traditional landfill and incineration methods. Finding revenue streams and cost recovery mechanisms to make these practices economically viable can be challenging and require cost-efficient solutions. Government subsidies could be a potential example to make these upfront sustainable investments more attractive and less expensive.
- Market Demand: Creating a robust market for recycled materials and compost is crucial for the success of Zero Waste, Circular Economy, and TRUE certification. Limited demand in specific sectors can discourage investment and slow down waste diversion efforts.
- Consumer Behavior: Shifting consumer behavior toward responsible waste disposal and resource conservation requires extensive awareness campaigns and educational programs to overcome ingrained habits, current cultural norms, and the status quo.

Policy and Regulatory

- Lack of Standardization: Clear and consistent regulations and standards for waste management practices, labeling for recycled materials and product end-of-life responsibility, are a few examples of crucial guidelines to help enable a seamless transition toward Zero Waste, TRUE, and Circular Economy. Gaps in these areas can ultimately delay the progress. Ensuring that governmental regulations support, facilitate, and progress the principles of Zero Waste and resource efficiency is crucial for success.
- Enforcement and Transparency: Implementing and enforcing stricter regulations on waste disposal, promoting, and maintaining transparency in waste management systems are essential to hold all stakeholders accountable and discourage practices that contradict and or prevent progress of the Zero Waste goals.

Social and Cultural

- Changing Mindsets: Overcoming traditional perceptions of waste as disposal and fostering a "resource recovery" mindset among all governments, companies, and individuals requires cultural shifts and promotion of the value of resource efficiency and circularity.
- Workforce Availability: Implementing and maintaining advanced waste management systems requires skilled personnel in various areas, potentially demanding upskilling or attracting specialized talent, which can be challenging in rapidly evolving and complex markets.
- Awareness and Education: Generating awareness and fostering behavioral changes among residents, businesses, industries, and governments is pivotal. Without a widespread understanding of the importance of waste reduction and recycling, achieving compliance with Zero Waste, TRUE standards, and Circular Economy most likely will remain challenging.

In addition to the above categories, specifically the IT industry faces additional challenges when it comes to TRUE certification in the UAE. For example:

- Specialized Waste Streams: Data centers produce unique waste, including electronic waste (e-waste) and specialized equipment (that is near impossible to recycle). Managing these streams effectively to meet TRUE Waste standards might require specialized recycling facilities and disposal methods; details, such as this still need to be finalized, communicated effectively, and then implemented.
- Technological Obsolescence: Rapid technological advancements often lead to frequent upgrades and replacements of "out of date" equipment in data centers and technology companies. Disposing of outdated technology hardware and infrastructure while adhering to Zero Waste and TRUE principles poses significant challenges in ensuring responsible disposal and recycling.
- Data Security Concerns: The destruction and or recycling of data storage devices carries inherent security risks. Balancing data security requirements with responsible disposal or repurposing of devices remains a challenge.
- Lack of Local Recycling Infrastructure: Specialized recycling facilities for electronic components are limited in the UAE. Dependence on external recycling

facilities and or shipping waste abroad, outside of the UAE, for processing would negatively affect the attempted achievement of TRUE goals.

- Regulatory Compliance: Ensuring that local regulations align with TRUE Waste standards for electronic waste disposal and recycling might require amendments or specific exemptions to accommodate the unique nature of data center waste.

It should be highlighted that:

> While plastic waste has garnered huge attention in recent years due to its impacts on marine life, electronic waste (e-waste) is the fastest growing waste stream worldwide. Global e-waste generation stood at around 50 million metric tons in 2019 and was forecast to grow by around 40 percent in just one decade, to reach some 75 million tons by 2030. Much like other waste forms, e-waste generation per capita is higher in wealthier nations, with countries like Norway, the UK, and Switzerland generating more than 20 kg per person per year. While e-waste is a fast-growing stream, food is still the most common form of waste, accounting for almost 50 percent of global MSW generation. (Alves, 2023a, 2023b, 2023c)

Is Zero Waste Failing?

It is not entirely accurate to say that Zero Waste is failing altogether. While achieving TRUE waste on a large scale remains a significant challenge, it is still a necessary and viable goal and numerous successes have been achieved. To date, many companies, cities, countries, and regions have achieved impressive waste diversion rates through investments in infrastructure, awareness campaigns, and policy changes. Technological advancements, innovation in recycling, composting, and waste-to-energy technologies offer promising solutions for the future. Numerous communities and organizations have achieved near Zero Waste goals, demonstrating the potential for wider adoption and implementation.

Therefore, while achieving absolute Zero Waste on a global scale is still a considerable challenge, it is not a failed nor failing movement. By acknowledging the obstacles, learning from existing successes, and continuously innovating, we can steadily move toward a more sustainable future with minimized waste generation and maximized resource recovery.

Recommendations

Providing Additional Recommendations for TRUE Adherence in the UAE

Addressing the various challenges of TRUE certification in the UAE requires a combination of innovative approaches and viable solutions:

- Policy Support: Aligning local regulations and policies with TRUE principles is crucial for widespread adoption. This involves integrating TRUE concepts into sustainability policies, guiding procurement practices, building designs, and even human resource policies to incentivize employee engagement in waste reduction. By aligning regulations and policies with TRUE standards, we establish an environment where sustainability becomes the default setting.
- Data Collection and Reporting: Accurate waste management data is vital for shaping local policies amid urbanization and population growth. Knowing waste quantity and types helps authorities select suitable strategies and prepare for future needs. Transparent reporting enhances accountability and provides stakeholders with clear insights, fostering informed decision-making and public awareness. As per the physicist Lord Kelvin "If you cannot measure it, you cannot improve it."
- Education and Awareness Campaigns: Extensive campaigns are essential to educate the public and stakeholders about the benefits of TRUE practices. Engaging videos and interactive workshops demonstrate recycling and composting benefits. Empowering stakeholders with knowledge and opportunities for input fosters support.
- Financial Support and Incentives: Providing financial incentives and support can promote TRUE standards adoption. Governments and organizations can offer grants to cover costs of equipment, training, or certification, encouraging businesses to view TRUE as an investment in sustainability. This benefits the environment, economy, and society overall.

Innovative Approaches

- Implement gamification for waste reduction: Develop engaging mobile apps and/or reward systems that incentivize individuals, companies, and communities to participate in waste reduction and recycling initiatives.

Conclusion

In conclusion, this paper emphasizes the importance of addressing waste generation, environmental degradation, and resource conservation. The statistics presented underscore the magnitude of the issue, with different regions facing varying levels of waste generation, posing a significant threat to the environment.

The United Arab Emirates has emerged as an active participant and thought leader in the global movement toward sustainable waste management. Local authorities and initiatives, such as Tadweer, the "Waste to Zero" initiative, and the UAE government's strategic policies, showcase the nation's commitment to align with international sustainability best-in-class practices.

The unique challenges faced by the UAE in pursuing Zero Waste Goals have been highlighted: from infrastructure limitations to cultural shifts and regulatory frameworks. It identifies the TRUE certification as a valuable tool for corporations in the UAE, and the challenges associated with the TRUE certification. The examination of these challenges has been carried out through site visits, waste audits, and gap analyses which provide a comprehensive understanding of the obstacles faced by organizations striving for TRUE certification.

Lastly, this paper proposes recommendations tailored to address the challenges specific to the UAE. These include policy support, data collection and reporting transparency, education and awareness campaigns, financial support, and incentives. Innovative approaches, such as gamification for waste reduction and financial incentives, are suggested to actively engage individuals and corporations.

Furthermore, this white paper provides a comprehensive overview of the waste management landscape, articulates the challenges faced in pursuing Zero Waste Goals in the UAE, and offers strategic solutions to accelerate progress toward a more sustainable and responsible waste management future.

Zero Waste is more of a guiding principle and an ambitious goal rather than a single solution. The journey toward TRUE is crucial, even if complete elimination of waste remains unlikely.

In conclusion, the intention of this paper is to investigate and analyze Waste Challenges when implementing TRUE certification. A further study should examine the data of waste production, segregation, and disposal by industry, in both the UAE and globally.

References

Alves, B. (2023, May 26). *Municipal waste generation by select country*. Statista. https://www.statista.com/statistics/916749/global-generation-of-municipal-solid-waste-by-country/

Alves, B. (2023, September 28). *Per capita municipal waste generation by country*. Statista. https://www.statista.com/statistics/1336513/global-generation-of-municipal-solid-waste-per-capitaby-country/#statisticContainer

Alves, B. (2023, December 18). *Topic: Waste generation worldwide*. Statista. https://www.statista.com/topics/4983/waste-generation-worldwide/#topicOverview

Andersen, I. (2022, October 03). *Towards a zero waste society: 30 years of UNEP-IETC*. UNEP. https://www.unep.org/news-and-stories/speech/towards-zero-waste-society-30-years-unep-ietc

Dubai Municipality. (2022, July 26). *Dubai Waste Management Centre to start initial operations in 2023*. https://www.dm.gov.ae/2022/07/26/dubai-waste-management-centre-to-start-initial-operations-in2023/

Emirates News Agency-WAM. (2022, January 27). *Executive Council of Dubai approves Integrated Waste Management Strategy 2021–2041, with a budget of AED74.5 billion*. https://wam.ae/en/details/1395303015567

Emirates News Agency-WAM. (2023, January 29). *UAE Circular Economy Council discusses 'year of sustainability' initiatives and COP28*. https://wam.ae/en/details/1395303123107

Emirates News Agency-WAM. (2019, October 27). *Tadweer to collaborate with Green Business Certification to achieve zero waste management*. https://wam.ae/en/details/1395302797769

It's time for a circular economy. How to Build a Circular Economy | Ellen MacArthur Foundation. https://www.ellenmacarthurfoundation.org/

Kaza, S., Yao, L., & Van Woerden, F. (2018, September 20). (rep.). *What a waste 2.0 a global snapshot of solid waste management to 2050*. World Bank. https://datatopics.worldbank.org/what-a-waste/trends_in_solid_waste_management.html

Maheshwari, A. (2022, October 20). *When waste is a resource.* https://www.synergyconsultingifa.com/industry-knowledge/when-waste-is-a-resource/#:~:text=More%20specifically%20the%20Middle%20East,expected%20to%20triple%20by%202050

Plockova, J. (2023, January 24). *Implementing zero waste at scale with true portfolios.* U.S. Green Building Council. (n.d.). https://www.usgbc.org/articles/implementing-zero-waste-scale-true-portfolios

Saleh, S. (2023, September 28). *UAE: Quantity of non-hazardous waste in Dubai by treatment method.* Statista. https://www.statista.com/statistics/1415849/dubai-quantity-of-non-hazardous-waste-by-treatment-method/

Tadweer. (2023, December 06). *Waste to zero initiative officially launched at COP28, H.E. Mariam bint Mohammed Almheiri Assumes Chairmanship.* https://www.tadweer.ae/media/Waste%20to%20Zero%20Initiative%20Officially%20Launched%20at%20COP28,%20H.E.%20Mariam%20bint%20Mohammed%20Almheiri%20Assumes%20Chairmanship.html

The Official Portal of the UAE Government. (2023, August 14). *UAE circular economy policy: UAE Circular Economy Policy | The Official Portal of the UAE Government.* https://u.ae/en/about-the-uae/strategies-initiatives-and-awards/policies/economy/uae-circular-economy-policy

Tiseo, I. (2023, December 6). *Largest waste producing countries per capita.* Statista. https://www.statista.com/statistics/1168066/largest-waste-producing-countries-worldwide-per-capita/#:~:text=Worldwide%2C%20the%20biggest%20producer%20of,this%20generated%20by%20industrial%20waste

Waste Department. Dubai Municipality. (2024, January 12). https://www.dm.gov.ae/municipality/business/waste-department-2/

White Paper: 2040 vision for a sustainable future requires radical rethink on EU materials and waste policy. Zero Waste Europe (2023, April 20). https://zerowasteeurope.eu/press-release/white-paper-2040-vision-for-a-sustainable-future-requires-radical-rethink-on-eu-materials-and-waste-policy/

World Bank. *Who we are.* https://www.worldbank.org/en/who-we-are

Zaman, S. (2022, June 8). Video: New initiative in UAE to reduce DH6 billion worth of food wastage every year. *UAE–Gulf News.* https://gulfnews.com/uae/video-new-initiative-in-uae-to-reduce-dh6- billion-worth-of-food-wastage-every-year-1.1654683455221

Zero Waste. (2021, January 5). The top 10 Zero waste companies Today! Zerowaste.com. https://www.zerowaste.com/blog/top-10-zero-waste-companies/

Zero Waste Definition. (2022, September 7). *Zero waste international alliance.* https://zwia.org/zerowaste-definition

Chapter 17
The Use of Environmental Product Declarations to Inform Decision-Making

Deidre Wolff and Dania Ramadan

Abstract The buildings and construction industry is a significant contributor to global CO_2 emissions. Accurate environmental impact data is required to enable decision-makers to implement measures to reduce these emissions. For this reason, the demand for product environmental information that allows for comparisons between products is increasing. By providing public access to aggregated product environmental impact data, Environmental Product Declarations (EPDs) are a step in the right direction. EPDs communicate the results of third-party verified product Life Cycle Assessment (LCA) studies for multiple environmental impact categories, including climate change, and are commonly used within building rating systems, such as LEED and Mostadam. However, there is still work to be done in terms of harmonization of the methods used in the underlying LCA studies of EPDs. While EPDs of the same product that have applied different methods are often incomparable, the data presented in EPDs enables decision-makers to make better-informed decisions that can contribute positively to broader sustainability goals. This work focuses on assessing the comparability of EPDs for steel and concrete products manufactured in the GCC region. Results suggest that for steel, the EPDs are mainly from the same Programme Operator (PO) and are comparable. Whereas for concrete, EPDs are from multiple POs and differ in key methodological choices. These EPDs should therefore be examined thoroughly before their application in LCAs for building rating systems to avoid ineffective decisions from being implemented. The harmonization of methods behind LCA studies is an ongoing challenge that spans beyond the buildings and construction industry. Until harmonization and consensus are achieved, transparent reporting on the appropriate application of EPDs can help ensure decisions lead to reduced potential environmental impacts.

Deidre Wolff and Dania Ramadan were affiliated with AESG at the time of the study. Deidre Wolff is now a freelance EPD verifier and LCA consultant. Dania Ramadan is now affiliated with VestinGrow.

D. Wolff (✉) · D. Ramadan
AESG, Dubai, UAE
e-mail: dwolff@epdverify.com

D. Ramadan
e-mail: rramadan.dania@gmail.com

© The Author(s), under exclusive license to Springer Nature Switzerland AG 2026
J. Dsilva (ed.), *Redefining Sustainability*, https://doi.org/10.1007/978-3-031-88376-7_17

Keywords Life cycle assessment · Environmental product declaration · LEED · Buildings and construction · Decision-making

Introduction

The buildings and construction industry is a significant contributor to global carbon dioxide (CO_2) emissions. Within the UAE, this industry contributed to approximately 27% of total greenhouse gas emissions in 2019 and is thus a key sector for implementation of decarbonization efforts to meet 2030 targets (Emirates Green Building Council [EGBC], 2024). Globally, the International Energy Agency (IEA) reported that the industry accounted for approximately 37% of global operational and process-related emissions in 2021 (United Nations Environment Programme [UNEP], 2022). Process-related emissions are emissions associated with the manufacturing process for a building material, such as concrete or steel. This includes the emissions from the extraction of raw materials and from the production of materials and energy used within the manufacturing process. Process-related emissions account for approximately 9% of the 37% of global operational and process-related emissions (UNEP, 2022) and can be quantified using Life Cycle Assessment (LCA).

LCA is the compilation and evaluation of the inputs, outputs, and the potential environmental impacts of a product system throughout its life cycle (International Organization for Standardization [ISO], 2006c). A variety of environmental impact categories can be assessed with LCA, including climate change, acidification, eutrophication, photochemical ozone formation, and resource depletion (metals/minerals), to name a few. For a holistic assessment, multiple impact categories should be evaluated to avoid burden-shifting from one impact category to another. In other words, if focus is placed on one impact category (such as climate change), implementation of reduction measures for this impact category may result in an increase in another impact category (such as resource depletion—metals and minerals). By including multiple impact categories in a study, trade-offs can be identified and avoided or optimized.

LCA has been used across multiple sectors to inform decision-making and policy development. There are international standards that should be complied with when conducting an LCA study, being ISO 14040:2006 (ISO, 2006b) and ISO 14044:2006 (ISO, 2006c). These standards are guidelines that present the steps of an LCA and the methodological choices. The flexibility of these standards allows for LCA to be applied across multiple industries; however, the standards are also often criticized for being too vague leading to inconsistent application of methods and results that are incomparable. For this reason, additional standards, such as ISO 14025:2006 (ISO, 2006a), ISO 21930:2017 (ISO, 2017), and EN 15804:2012 (European Standard [EN], 2020), have been developed with the aim of harmonizing the methodological choices for product categories. The intent of harmonization is to enable valid comparisons within product categories.

ISO 14025:2006 provides the procedures for developing Environmental Product Declarations (EPDs) as well as guidance on the development of product category rules (PCRs). An EPD communicates the results of third-party verified product LCA studies, whereas PCRs provide the methods that the LCA study must comply with. ISO 21930:2017 and EN 15804:2012 provide the core rules for EPDs of construction products and services. However, in attempts to further harmonize the methods applied, additional standards, referred to as complementary product category rules (c-PCR), have also been developed that are product specific. For example, c-PCRs for concrete and concrete elements (EN 16757:2022) and for concrete batching plants have been published by the International EPD System.

Within the buildings and construction industry, EPDs are commonly used to communicate product LCA results, and the demand for more published EPDs is increasing. According to Gelowitz and McArthur (2017), the release of LEED version 4 has resulted in the initial increase in the number of EPDs published for the construction sector, as well as an increase in Programme Operators (POs) and developed PCRs and c-PCRs. POs are responsible for the verification and publication process of EPDs and may also publish their own PCRs and c-PCRs. International rating systems, such as LEED, have been used in the UAE since 2006 with approximately 386 certified projects as of mid-2020 (EGBC, 2020) and steadily growing. However, a key challenge of the use of EPDs within rating systems is comparability. In order to correctly determine if one product has a lower environmental impact over another, EPDs of similar products must be comparable. With the variation in methods applied, this is not always the case, which may lead to the selection of a product that does not necessarily have a reduced impact.

Work is being done to further harmonize the methodologies behind the LCA studies of EPDs in order to ensure comparability. One recommendation is to move toward one centralized PO (Konradsen et al., 2024). While this could be one part of the solution, the ability to come to a consensus on how to conduct an LCA study has a history of not being a simple task. While EPDs do outline the main methods that have been followed along with a disclaimer that EPDs from different POs may not be comparable, more is required to ensure their appropriate use.

Recent studies (Konradsen et al., 2024; Moré et al., 2022) have concluded that EPDs are largely incomparable due to the application of incomparable methods; however, as far as is known, a study within the GCC region has not been conducted. The purpose of this study is to conduct a methodological evaluation of the EPDs published for steel and concrete manufactured in the GCC region to determine their comparability and the implications on the use of these EPDs within regional building LCA studies conducted to obtain credits within rating systems, such as LEED.

Methodology

Available EPDs from steel (reinforcement steel) and concrete manufacturers in the GCC region were collected through a search of EPD libraries, as well as the ECO Platform (2024). The initial search resulted in over 70 EPDs for concrete and 8 EPDs for reinforcement steel. The concrete datasets were further filtered to remove duplicate datasets and to focus on ready-mix concrete of a similar grade. The main reason being that the use of these EPDs within LCAs conducted as part of building rating certification systems is often done based on material type and the associated environmental impact, and with a focus on structural components. The aim was to determine whether these EPDs for similar products could be considered comparable.

For the evaluation of the comparability, the method applied followed the methodology developed by Gelowitz and McArthur (2017) and further applied by Moré et al. (2022). With this methodology, a detailed examination of the EPDs across defined critical and crucial categories was conducted. Factors, such as PCR followed, functional unit, system boundary, allocation method, and data quality (specific versus generic), were critically examined, recognizing that variations in data sources, methodologies, and boundary conditions may impact comparability. All data used within the assessment was gathered from the information provided within the EPDs.

Findings

After filtering the EPDs collected for reinforcement steel and concrete (Grade C35-C50), a comparative evaluation was conducted for 8 reinforcement steel EPDs and another for 9 ready-mix concrete EPDs.

It was found that for reinforcement steel within the region, EPDs were published from the same PO and largely applied the same methodological assumptions, including allocation method, background database, end-of-life scenarios, and functional unit. The variation in these datasets was mainly found to be due to the use of different PCRs and underlying standards. The main reason for this was found to be the transition from standard EN 15804:A1 to EN 15804:A2. The differences between these standards include the subdivision of Global Warming Potential into three categories and a parallel subdivision of Eutrophication into three categories. Additionally, there are optional impact categories, and modules C1–C4 and D are mandated in EN 15804:A2. Because of this, it was found that caution should be used when comparing EPDs from steel manufacturers, particularly for the C and D modules. However, as EN 15804:A1 is phased out, comparability is likely to increase as long as similar methods continue to be applied.

For the ready-mix concrete EPDs, the results varied. In this case, multiple POs were used for the verification and publication of the EPDs. It was found that EPDs from different POs differed in the methodology applied, particularly regarding allocation method and system boundary cutoff rules. It was further found that there was variation in the PCR followed, resulting in differences in system boundary and impact category measurement units (some of which are not convertible). Because of this, it is recommended that the EPDs for concrete are thoroughly examined for their comparability before decisions are made based solely on the reported EPD results.

When considering the use of these EPDs within building rating systems, the reinforcement steel EPDs can be considered relatively comparable, and therefore, the implications for their use to reduce the environmental impact are less likely to be incorrect. However, more EPDs are required to verify this, and caution should be taken if comparing a steel EPD within the GCC region to another EPD published in a different region and by a different PO. For concrete, however, due to the variation in methodology, implications of replacing one EPD with another without considering the methodology followed have the potential to unknowingly increase as opposed to decrease the overall environmental impact.

Conclusions

The work presented in this paper focused on assessing the comparability of EPDs for steel and concrete products available from manufacturers in the GCC region. By evaluating two structural products, the implications for the use of these EPDs within building rating certification systems could be examined.

In general, it was found that the possibility of using one common PO has resulted in more comparable EPDs—meaning that harmonization of methods aids comparability. However, more EPDs are required to further verify the results obtained. Particularly, as more POs come onto the market and the demand for EPDs continues to rise, the importance of harmonization of methodologies applied across POs will be further exposed.

The ability to compare one EPD to another requires knowledge of LCA and the implications of varying assumptions; therefore, caution should be taken if solely comparing the result of one EPD to another. It was found that particularly for the concrete EPDs evaluated, key methodological differences resulted in incomparable EPDs. Therefore, the direct replacement of one EPD with another within building rating systems may lead to unknowingly increasing as opposed to decreasing the overall potential environmental impact.

Further research is required to verify the results obtained and provide more concrete solutions for methodology harmonization across EPDs. Research should focus on increasing the number of product categories and manufacturers across the GCC region that have been evaluated within this study. Increasing the number of EPDs assessed will further reveal methodological inconsistencies that lead to incomparability. It is expected that the number of published EPDs for the region will increase in the near future and therefore work to improve the comparability of EPDs will help ensure they can be effectively used in decision-making throughout the early design stages of buildings and construction projects and within LCA studies for building rating systems.

Research is continuing in the area of methodology harmonization to reach a consensus within many industries; thus, it is not only a challenge for the buildings and construction industry. In the meantime, transparency in reporting LCA results and methods within EPDs in an understandable way to non-LCA experts can also help ensure the appropriate use of EPD data. By outlining how the EPD results can be used, consumers and end-users are better equipped to identify cases where EPDs may or may not be comparable. With this, clarity is provided, and EPDs can move further toward accomplishing their initial aim of permitting comparability within product categories and being a useful tool for decision-making.

References

ECO Platform. (2024). *ECO Portal*. https://www.eco-platform.org/epd-data.html

Emirates Green Building Council. (2020). *Green building market brief—United Arab Emirates*. https://emiratesgbc.org/wp-content/uploads/2020/09/UAE-Brief_NewTemplate_electronic_final.pdf

Emirates Green Building Council. (2024). *UAE sustainability built environment blueprint preview—United Arab Emirates*. https://emiratesgbc.org/wp-content/uploads/2023/12/EDG_EGBC_UAE-Sustainability-Built-Environment-Blueprint_Preview-1.pdf

European Standards. (2020). *Sustainability of construction works—Environmental product declarations—Core rules for the product category of construction products*. EN Standard No. 15804:2012+A2:2020. https://www.en-standard.eu/une-en-15804-2012-a2-2020-sustainability-of-construction-works-environmental-product-declarations-core-rules-for-the-product-category-of-construction-products/

Gelowitz, M. D. C., & McArthur, J. J. (2017). Comparison of type III environmental product declarations for construction products: Material sourcing and harmonization evaluation. *Journal of Cleaner Production, 157*, 125–133. https://doi.org/10.1016/j.jclepro.2017.04.133

International Organization for Standardization. (2006a). *Environmental labels and declarations—Type III environmental declarations—Principles and procedures* (ISO Standard No. 14025:2006). https://www.iso.org/standard/38131.html

International Organization for Standardization. (2006b). *Environmental management—Life cycle assessment—Principles and framework* (ISO Standard No. 14040:2006). https://www.iso.org/standard/37456.html

International Organization for Standardization. (2006c). *Environmental management—Life cycle assessment—Requirements and guidelines* (ISO Standard No. 14044:2006). https://www.iso.org/standard/38498.html

International Organization for Standardization. (2017). *Sustainability in buildings and civil engineering works—Core rules for environmental product declarations of construction products and services* (ISO Standard No. 21930:2017). https://www.iso.org/standard/61694.html

Konradsen, F., Hansen, K. S. H., Ghose, A., & Pizzol, M. (2024). Same product, different score: How methodological differences affect EPD results. *International Journal of Life Cycle Assessment, 29*, 291–307. https://doi.org/10.1007/s11367-023-02246-x

Moré, F. B., Galindro, B. M., & Soares, S. R. (2022). Assessing the completeness and comparability of environmental product declarations. *Journal of Cleaner Production, 375*, 133999. https://doi.org/10.1016/j.jclepro.2022.133999

United Nations Environment Programme. (2022). *2022 global status report for buildings and construction: Towards a zero-emission, efficient and resilient buildings and construction sector*. https://www.unep.org/resources/publication/2022-global-status-report-buildings-and-construction

Part III
Resilient Cities 2050 and Sustainable Environment

Chapter 18
Evaluating Visual Comfort Metrics in Dynamic Daylit Spaces: Challenges and Opportunities

Dalia Hafiz

Abstract "Natural light is the only light because it has mood… it puts us in touch with the eternal. It is the only light that makes architecture."—Louis Khan (Baker and Steemers, Daylight design of buildings: a handbook for architects and engineers 1–250, 2014; Tango, Object in light: from power to purpose, 2011). Daylight is one key aspect that enhances the sense of place and influences the personal interpretations and impressions that last long after leaving the place. However, visual discomfort and glare can distract architects from maximizing daylighting. To better achieve visual comfort in dynamic daylit spaces, the time and space dynamics of daylight conditions and the representation and re-imagining of these dynamics need to be considered. This paper examines logical argumentation through the exploration of various visual comfort metrics. The argument investigates several factors that affect visual performance and occupants' comfort. Some of these factors include lighting level, uniformity of illuminance, color rendering, avoiding hard shadows, contrast rendition, physiological glare, balanced brightness distribution, luminance levels variation, discomfort glare, illuminance uniformity in the area around the visual task, and the balance between artificial lighting and daylight. The argumentation concludes with a compilation of different visual comfort evaluation metrics. These metrics aim to consider the two fundamental dynamics of daylighting, time and movement through spaces, to achieve adequate daylighting performance with the most visual comfort. The research outcomes are expected to provide researchers, designers, and decision-makers with a new approach to designing to improve visual comfort, energy performance, and the quality of the place.

Keywords Occupant comfort · Daylighting · Visual comfort · Visual comfort metrics · Glare metrics

---

D. Hafiz (✉)
Ajman University, Ajman, UAE
e-mail: d.hafiz@ajman.ac.ae

© The Author(s), under exclusive license to Springer Nature Switzerland AG 2026
J. Dsilva (ed.), *Redefining Sustainability*, https://doi.org/10.1007/978-3-031-88376-7_18

Introduction

Arguments are chains of reasoning that lead to a conclusion while considering potential counterarguments at each step. They can present knowledge and common-sense reasoning. The foundation of a logical argument lies in its intention, where the proposition is either true or false (i.e., accurate or inaccurate). For an argument to be rational and valid, it must follow three stages: premises, inference, and conclusion.

A. Premises: These are the necessary propositions for the argument to progress. They are indicated by phrases, such as "because," "since," or "obviously".
B. Inference is the process of using the premises to derive further propositions. It is typically indicated by phrases, such as "implies that" or "therefore".
C. Conclusion: This is the final stage of inference, affirming the argument based on the premises and the inferences drawn. Conclusions are often indicated by phrases, such as "therefore," "it follows that," or "we conclude".

In this research, the primary goal of logical argumentation is to advocate for selecting thresholds for design criteria and decision support metrics related to visual comfort. The paper is structured according to the stages of argumentation:

- Identification of the argumentation method,
- Presentation and categorization of evaluation metrics related to glare and visual comfort thresholds,
- Argumentation for the selection of each metric and its applied threshold,
- Conclusion of the argumentations,
- Testing of the argument findings and conclusions through a case study of a typical office space where the selected metrics and their thresholds are applied,
- Conclusions from the case study, along with contributions to design decision-making processes,
- Argumentations concerning visual comfort metrics in dynamic daylit spaces can address challenges in the built environment by:
  - Enhancing daylight exposure's psychological and physiological benefits, such as reducing stress, improving mood, and regulating circadian rhythms.
  - Optimizing natural light while minimizing unwanted heat gain or loss can reduce the energy consumption and carbon footprint associated with artificial lighting, heating, and cooling systems.
  - Improving the visual performance and productivity of building occupants by ensuring adequate and uniform illumination levels, avoiding glare and contrast issues, and providing views to the outside, thus contributing to a more comfortable and productive work environment.
  - Increasing spaces' aesthetic and architectural value through dynamic and diverse lighting effects, enhancing spatial perception and ambiance, and expressing the design intent and identity.

Visual comfort metrics in dynamic daylit spaces aim to investigate their performance and validity across various spaces and conditions and propose new or improved metrics to predict better and prevent visual discomfort and glare. This topic is essential as it can aid designers and architects in creating more comfortable and sustainable spaces that benefit both occupants and the environment.

Several factors influence visual performance, including lighting levels, illuminance uniformity, color rendering, avoidance of harsh shadows, contrast rendition, physiological glare, balanced brightness distribution, and variations in lighting conditions.

Methodology

Several factors affect visual performance, including lighting level, illuminance uniformity, color rendering, avoiding hard shadows, contrast rendition, physiological glare, balanced brightness distribution, variation of luminance levels, and discomfort glare. The literature found that researchers cannot depend on a single metric to investigate visual discomfort aspects, Fig. 18.1. Moreover, most of the indices examined in previous research were devoted to predicting light distribution, glare phenomena, and light quality (Cipriano et al., 2015); consequently, metrics representing these different aspects were used (Building 2022, n.d.; Mentens et al., 2021a).

Regarding the visual comfort evaluation metrics argument, the useful daylight illuminance (UDI) and illuminance value distribution were used to evaluate the light distribution in the space. For light quality, the luminance ratio (RML) was used. Finally, the Daylight Glare Probability (DGP) and the Daylight Glare Index (DGI) were used for glare evaluation metrics, as argued in the following sections (Chaloeytoy et al., 2020; Mentens et al., 2021a).

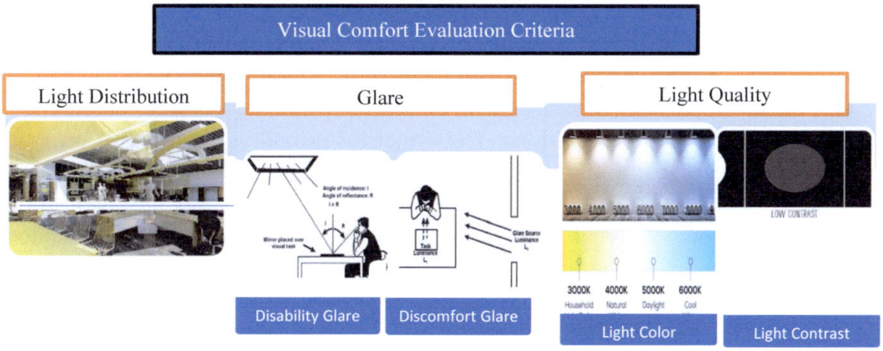

Fig. 18.1 Visual discomfort examined aspects

Light Distribution: Illuminance-Based Metrics

Short execution times are essential for providing timely feedback and decision-making support to designers in the early stages of the design process. Since illuminance simulations require less computational time than luminance image simulations, illuminance metrics were used to evaluate visual comfort.

Useful Daylight Illuminance (UDI)

The useful daylight illuminance (UDI) metric can effectively measure both over-lit and under-lit conditions, providing valuable insights into light distribution within a space. As a result, UDI is suitable for both short-term and long-term evaluations, particularly in situations where excessive sunlight can lead to intense glare. Since UDI calculations are based on spatial renderings for each point in the calculation grid, it can ensure that all simulation points comply with the established illuminance guidelines, as suggested by Yin. In this research, UDI was tested on the solstices and equinox days only (Brzezicki & Regucki, 2020; Brzezicki 2021; Yılmaz 2022).

Glare: Luminance-Based Metrics

The luminance evaluation focused on spatial and temporal instances corresponding to peak illuminance conditions. While the current analysis is constrained to these peak periods due to limitations in computational processing speeds, future technological advancements are expected to facilitate more comprehensive hourly luminance calculations. Such detailed temporal analyses would better understand daylight dynamics and their implications on visual comfort.

One or more established metrics with defined threshold levels were employed to effectively identify glare conditions. These metrics provide a quantitative basis for evaluating visual discomfort and glare potential. Subsequent sections delve into the specific luminance metrics utilized in this study, offering insights into their application and relevance to dynamic daylit environments (Cipriano et al. n.d.; Mentens et al., 2021b).

Daylight Glare Probability (DGP)

Comfort is fundamentally defined by user satisfaction, with glare being a subjective phenomenon that varies significantly among individuals. Given this variability, employing a probability distribution provides a robust approach to predicting comfort

Table 18.1 DGI thresholds

DGI effect	Imperceptible	Perceptible	Disturbing	Intolerable
Value	< 18	18–24	24–31	> 31
Normalized value (*0.01452)	< 0.26136	0.26136–0.34848	0.34848–0.45012	> 0.45012

levels. Previous studies have demonstrated a strong correlation between the Daylight Glare Probability (DGP) metric and occupant satisfaction, highlighting its efficacy as a predictive tool for assessing visual comfort. Consequently, the DGP was adopted as the primary visual comfort indicator in the prototype, ensuring that the assessment aligns with established research findings and occupant-centric considerations (Chaloeytoy et al., 2020).

Daylight Glare Index (DGI)

The Daylight Glare Index (DGI) is a widely used metric for assessing visual comfort, particularly in spaces with significant glare sources, such as windows with direct sky exposure. The DGI quantifies the level of discomfort caused by glare, considering the glare source's luminance relative to the surrounding field of view (Jakubiec & Reinhart, 2011). The DGI can be efficiently calculated using tools, such as Evalglare, a software program that processes high-dynamic-range (HDR) images. Evalglare integrates key variables, including luminance distribution, date, time, and prevailing sky conditions (Nabil & Mardaljevic, 2006). Additionally, it factors in the material properties of interior finishes, which can significantly influence glare perception. One notable advantage of the DGI is its ability to be normalized and compared with the average Daylight Glare Probability (DGP). This facilitates a more comprehensive evaluation of glare over time and across varying conditions, enabling architects and lighting designers to optimize daylighting strategies while minimizing discomfort, as shown in Table 18.1 (Jakubiec & Reinhart, 2011), It was used in the prototype for a more accessible representation of the glare condition.

Light Quality: Luminance-Based Metrics

The interaction between light color and contrast plays a pivotal role in determining overall light quality. While most light quality indices, such as the Color Rendering Index (CRI) and the Color Preference Index (CPI), primarily address artificial lighting, natural daylight remains the benchmark for achieving superior and inspiring lighting conditions. These metrics provide critical insights into how light influences our environment and inform strategies to optimize lighting design.

Recognizing and leveraging the positive effects of high-quality lighting is essential for enhancing both environmental aesthetics and human well-being (Hautière et al., 2008; Thornton, 1972; Xu, 1984). Image contrast was used as a metric for daylight quality evaluation.

Luminance Ratio

Previous research identified several luminance ratio guidelines applied to the three zones of the visual field (central, adjacent, and non-adjacent). This includes luminance ratios of 1:3:10 identified by Osterhaus (2002). Other research claims that occupants can tolerate a ratio of 1:40 and up to 1:100 between the central zone and the surroundings and a ratio of 1:3 for visual tasks and immediate surroundings. This research used the ratio of 1:3:10 as a threshold value (Yılmaz 2022; Linney, 2008).

Used Metrics Argument Conclusion

This study employed multiple evaluation metrics to assess visual comfort and daylighting performance. Useful daylight illuminance (UDI) and illuminance distribution were utilized for illuminance evaluation. Luminance image analysis included metrics, such as Daylight Glare Probability (DGP), Daylight Glare Index (DGI), and luminance ratio. Additionally, visual renderings were developed to represent and reinterpret the analyzed space, providing a comprehensive understanding of the visual environment, as applied in the case study investigated in this research.

Case Study Explanation

A case study was conducted to test the validity and effectiveness of the proposed arguments. The selected case study involved a typical side-lit office space, modeled as a single office measuring 15 ft × 9 ft (4.6 m × 2.7 m) with a west-facing window measuring 6 ft × 3 ft (1.8 m × 0.9 m) and a 3-foot-wide (0.9 m) corridor. A case study was conducted to test the validity and effectiveness of the proposed arguments. The selected case study involved a typical side-lit office space, modeled as a single office measuring 15 ft × 9 ft (4.6 m × 2.7 m) with a west-facing window measuring 6 ft × 3 ft (1.8 m × 0.9 m) and a 3-foot-wide (0.9 m) corridor (Fig. 18.2).

The 3D model incorporated various inputs, including geographical location, sky conditions, material properties, building geometry, and simulation points, summarized in Table 18.2.

Material properties for the 3D model were specified based on location. Generic radiance material properties were applied, as detailed in Table 18.3:

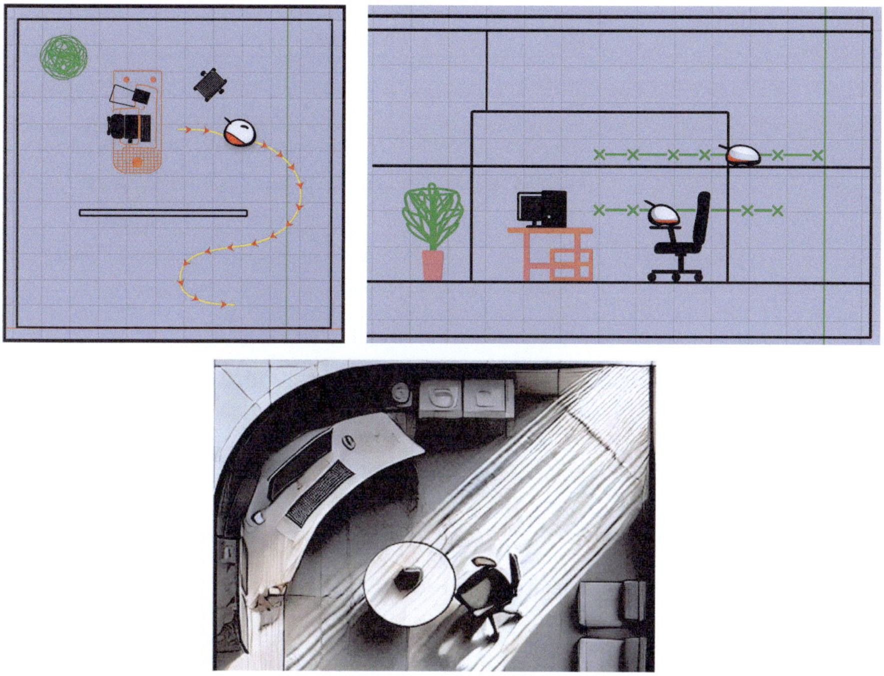

Fig. 18.2 Simulation points and layout

Table 18.2 Case study data inputs

Variable	Examined case properties
Geographical location	San Francisco weather file
Sky condition	Clear sky without sun; clear sky with sun
Date and time (Mm Dd)	Default annual (hourly intervals)
Material properties	Generic materials (detailed in Table 18.3)
Building orientation	Rotated to match Rhino's default orientation
Building geometry	Modeled in Rhino
Surroundings	No surrounding structures included
Simulation points	Placed along the circulation corridor in the 3D mode

Simulation Setup

Simulations were conducted at two eye levels to account for different occupant perspectives:

Standing/walking condition: Simulation points were placed horizontally along the circulation path at 4.5-foot (1.4 m) intervals and vertically at 5.5 feet (1.8 m).

Table 18.3 Materials properties

Item	Material
Ceiling	Generic ceiling (80% reflectance)
Interior floor	Generic floor (20% reflectance)
Interior walls	Generic interior wall (50% reflectance)
Outside façade	Generic façade (35% reflectance)
Outside ground	Generic ground (20% reflectance)
Glass	Single pane clear (90% transmittance)
Window frame	Gray metal (50% reflectance)

Sitting/working condition: Simulation points were positioned at average eye level for seated occupants, horizontally at 3.3 feet (1 m) and vertically spaced at 4.5 feet (1.4 m). Figure 18.2 illustrates the simulation points in plan and section views.

Evaluation Metrics

The evaluation process categorized visual comfort metrics into two groups:

Illuminance-based metrics: The assessed light distribution, including useful daylight illuminance (UDI) and hourly illuminance distribution on peak days.

Luminance-based metrics: These focused on glare evaluation, using Daylight Glare Index (DGI), Daylight Glare Probability (DGP), and luminance ratio to address light quality and contrast issues.

Illuminance simulations were conducted at fixed points along the path, followed by luminance image simulations to evaluate glare and overall visual comfort.

Comparison of Glare Metrics

The comparison of glare metrics highlights the relationship between useful daylight illuminance (UDI) and glare thresholds measured using Daylight Glare Probability (DGP) and Daylight Glare Index (DGI). The chart illustrates the performance of each metric, showing how UDI levels correlate with visual comfort thresholds for DGI and DGP. Threshold lines for DGI and DGP are depicted, providing a clear benchmark for acceptable glare levels. This comparison aids in understanding the effectiveness and limitations of each metric in evaluating glare within daylit spaces, as shown in Fig. 18.3 and explained below.

Glare Metrics

DGI (Normalized): The blue bar represents normalized DGI values. The value is around 0.3, indicating moderate glare under the "base case" scenario.

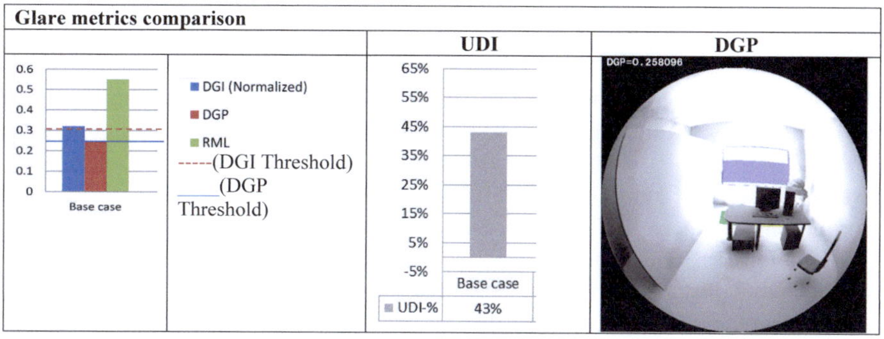

Fig. 18.3 Glare metrics comparison

DGP: The red bar shows the DGP, slightly above the threshold (around 0.3), indicating potential glare discomfort.

RML: The green bar represents RML, significantly higher than the other metrics. This suggests localized high luminance sources, potentially causing discomfort glare.

Threshold Line

The red dashed line indicates a threshold level (likely the discomfort glare threshold). Both DGI and DGP are near or slightly exceed the threshold, emphasizing glare concerns.

Useful Daylight Illuminance (UDI-%)

The UDI-% graph shows a value of 43% for the base case, reflecting the percentage of time the illuminance falls within the range considered useful for visual tasks (e.g., 100–2000 lx). A UDI-% of 43% suggests moderate daylight performance, with room for improvement.

Fisheye Image Analysis

The fisheye image indicates areas of high brightness (e.g., window or screen) contributing to glare. The DGP value of 0.258096 is slightly below the threshold for significant discomfort glare, but it might still cause mild discomfort depending on user sensitivity and tasks performed in the space.

Findings

The metrics highlight a moderate glare issue in the "base case" scenario, with localized sources (as seen in the RML and fisheye image) contributing significantly.

While DGP remains slightly below discomfort glare thresholds, the RML suggests possible hotspots of glare. The UDI-% also indicates room for optimizing daylight distribution to reduce glare and improve visual comfort.

Mitigation strategies, such as external shading devices, glare-reducing window films, or adjusting the workstation's layout, might help reduce glare while maintaining acceptable daylight performance.

Further simulations or field measurements could refine the analysis to explore alternative daylighting designs.

Some of the questions that this research can help answer include:

- How to design buildings that optimize the use of daylight and reduce the need for artificial lighting, which can save energy and reduce greenhouse gas emissions.
- How to balance the benefits of daylight, such as providing a connection to the outdoors and enhancing the mood and alertness of the occupants, with the potential drawbacks, such as glare, overheating, and uneven distribution of light.
- How to select and apply the most appropriate and reliable visual comfort metrics that can predict the subjective perception and preference of the occupants in different daylight conditions and space types.
- How to incorporate the feedback and adaptation of the occupants into the visual comfort assessment, as they may adjust their posture, position, or shading devices to improve their comfort.
- How to integrate the visual comfort metrics with other aspects of building performance, such as thermal comfort, indoor air quality, acoustics, and aesthetics.

Recommendations to Enhance the Space Based on Case Study Findings

- Implement External Shading Devices: Use adjustable or fixed shading elements to reduce direct sunlight and mitigate glare.
- Apply Glare-Reducing Window Films: Use advanced window films to decrease glare while allowing natural light.
- Optimize Workstation Layout: Adjust workstation positions to avoid direct sunlight or glare hotspots.
- Enhance Daylight Distribution with Reflective Surfaces: Use reflective materials to distribute daylight evenly and minimize glare hotspots.
- Fine-tune Daylighting Systems: Integrate automated systems like dimmable lights or dynamic shading solutions.
- Improve Visual Comfort Metrics: Use a variety of visual comfort metrics to evaluate glare and daylighting comprehensively.

- Conduct Further Simulations and Field Measurements: Perform dynamic simulations and field studies to refine glare assessments.
- Refine Design with Immersive Technologies: Use VR to simulate and test daylighting and glare mitigation strategies.
- Integrate AI and Machine Learning: Use AI tools to optimize daylighting designs and automate shading system adjustments.
- Balance Visual, Thermal, and Acoustic Comfort: Integrate visual comfort strategies with thermal and acoustic performance.
- Implement Adaptive Features for Occupant Feedback: Allow occupants to adjust their workspace and environmental settings.
- Explore Parametric Design for Dynamic Daylighting: Use parametric design to explore daylighting and shading solutions.
- Expand Research into Diverse Climates and Building Types: Conduct simulations across different climates and building types for comprehensive benchmarks.

Conclusion and Future Research

This study's findings confirm the methodology's effectiveness in comparing and evaluating glare and visual comfort metrics to identify the most suitable approaches. Although no single metric was determined to significantly detect glare more effectively than others, using multiple luminance-, illuminance-, and image-based metrics proved efficient and comprehensive in detecting glare. This multifaceted approach offers valuable insights that can positively influence design decision-making and enhance existing design practices. The research provided a nuanced understanding of glare and visual discomfort phenomena, emphasizing their role in shaping spatial dynamics and supporting theoretical advancements in design by integrating glare and visual comfort considerations into practice.

However, challenges remain in studying and applying visual comfort metrics in dynamic daylit spaces. These challenges include the lack of standardization in the definitions, calculations, and thresholds for various metrics and the complexity of predicting and simulating the dynamic performance of daylighting systems and their interaction with occupants and the environment. Balancing visual comfort with other aspects of building performance, such as thermal and acoustic comfort and indoor air quality, adds to this complexity. Future research will address these limitations by exploring parametric design approaches and conducting annual simulations to assess metrics under diverse conditions. Additionally, integrating virtual reality (VR) for real-time immersive experiences can enhance both qualitative and quantitative evaluations of glare and visual comfort, ultimately contributing to improved occupant comfort.

Future research will focus on refining glare detection and visual comfort assessment through parametric design approaches and dynamic simulations. These methods will better evaluate metrics under varying spatial and temporal conditions, optimizing daylight usage while mitigating glare and improving thermal and visual comfort.

Expanding annual performance assessments across diverse climates and architectural contexts can establish more reliable benchmarks for visual comfort. Immersive technologies like virtual reality (VR) will enhance evaluations by simulating realistic environments and capturing occupant feedback, providing insights into adaptive behaviors, such as posture adjustments or shading device use.

Integrating artificial intelligence (AI) into this research offers transformative possibilities. AI can analyze vast datasets from simulations and real-world studies, identifying patterns and generating predictive models to optimize daylighting designs. Machine learning algorithms can recommend the best combinations of visual comfort strategies based on building characteristics and occupant preferences. Furthermore, AI-powered tools can automate the design and evaluation, offering real-time adjustments to shading systems or lighting layouts for optimal comfort and energy efficiency. These advancements will contribute to more sustainable and occupant-friendly building designs.

References

Baker, N., & Steemers, K. (2014). Daylight design of buildings: A handbook for architects and engineers. In *Daylight design of buildings: A handbook for architects and engineers* (pp. 1–250). https://doi.org/10.4324/9781315073750

Brzezicki, M., & Regucki, P. (2020). Optimization of useful daylight illuminance vs. drag force for vertical shading fins/panels. *Science and Technology for the Built Environment*, *27*(3), 367–376. https://doi.org/10.1080/23744731.2020.1858679

Yılmaz, F. Ş. (2022, n.d.). Evaluation of daylight provision and daylight glare probability for diverse workspace configurations. *Meridian.Allenpress.Com.* Retrieved January 24, 2023, from https://meridian.allenpress.com/jgb/article-abstract/17/1/199/479536

Brzezicki, M. (2021). An evaluation of useful daylight illuminance in an office room with a light shelf and translucent ceiling at 51° N. *Mdpi.Com.* https://doi.org/10.3390/buildings11110494

Chaloeytoy, K., Ichinose, M., Buildings, S. C. (2020, n.d.). Determination of the simplified daylight glare probability (DGPs) criteria for daylit office spaces in Thailand. *Mdpi.Com.* https://doi.org/10.3390/buildings10100180

Cipriano, G., Fanizza, C., Cosentino, A., & Carlucci, R. (n.d.). *Carlucci et al 2015.* Retrieved December 26, 2024, from https://www.academia.edu/22468382/Carlucci_et_al_2015

Hautière, N., Tarel, J. P., Aubert, D., & Dumont, É. (2008). Blind contrast enhancement assessment by gradient rationing at visible edges. *Image Analysis and Stereology, 27*(2), 87–95. https://doi.org/10.5566/IAS.V27.P87-95

Jakubiec, J. A., & Reinhart, C. (2011, November). DIVA 2.0: Integrating daylight and thermal simulations using Rhinoceros 3D, Daysim and EnergyPlus. In *Building simulation 2011* (Vol. 12, pp. 2202–2209). IBPSA.

Linney, A. S. (2008). *Maximum luminance and luminance ratios and their impact on user experience.* https://www.bing.com/search?q=Linney%2C+A.+S.+(2008).+Maximum+luminances+and+luminance+ratios+and+their+impact+on+user+experience.+%5BJournal+Name%5D.&cvid=45f95b44a152460f966a712b22508a24&gs_lcrp=EgRlZGdlKgYIABBFGDkyBggAEEUYOTIICAEQ6QcY_FXSAQc1NjRqMGo5qAIIsAIB&FORM=ANAB01&PC=U531

Mentens, A. , Martin, S. , Descamps, F. , Lataire, J. , & Jacobs, V. A. (2021a). Daylight glare probability prediction for an office room. *Cris.Vub.Be.* https://doi.org/10.25039/x48.2021.OP20

Mentens, A. , Martin, S. , Descamps, F. , Lataire, J. , & Jacobs, V. A. (2021b). *Daylight glare probability prediction for an office room* (pp. 176–181). https://doi.org/10.25039/x48.2021.OP20

Nabil, A., & Mardaljevic, J. (2005). Useful daylight illuminance: A new paradigm for assessing daylight in buildings. *Lighting Research and Technology, 37*(1), 41–59. https://doi.org/10.1191/1365782805li128oa

Osterhaus, W. (2002). *Recommended luminance ratios and their application in the design of visual environments.* https://www.bing.com/search?q=Osterhaus%2C+W.+%282002%29.+Recommended+luminance+ratios+and+their+application+in+the+design+of+visual+environments.&qs=n&form=QBRE&sp=-1&ghc=1&lq=1&pq=osterhaus%2C+w.+%282002%29.+recommended+luminance+ratios+and+their+application+in+the+design+of+visual+environments.&sc=1-110&sk=&cvid=91912F158F6B480885E63ADB9970DC84&ghsh=0&ghacc=0&ghpl=

Tango, R. W. (2011). Object in light: From power to purpose. *Ijme.Us.* https://ijme.us/cd_11/PDF/Paper%20220%20INT%20302.pdf

Thornton, W. A. (1972). Color-discrimination index. *Journal of the Optical Society of America, 62*(2), 191–194. https://doi.org/10.1364/JOSA.62.000191

Xu, H. (1984). Color rendering capacity of illumination. *Journal of the Illuminating Engineer.* https://www.bing.com/search?q=Xu%2C+H.+%281984%29.+Color+rendering+capacity+of+illumination.+Journal+of+the+Illuminating+Engineer.+&qs=n&form=QBRE&sp=-1&lq=1&pq=xu%2C+h.+%281984%29.+color+rendering+capacity+of+illumination.+journal+of+the+illuminating+engineer.+&sc=0-95&sk=&cvid=4E3A183458BE4B058C7EDE1AA455959D&ghsh=0&ghacc=0&ghpl=

Chapter 19
An All Consequences Calculator for Roadworks and Streetworks Design

Kelwalee Jutipanya, Christopher D. F. Rogers, and Emma J. S. Ferranti

Abstract Streetworks and roadworks play a significant role in the emission of carbon dioxide and consequential climate change (Local Council Roads Innovation Group, 2023). Achieving Net Zero in this sector requires the implementation of innovative materials, equipment and processes to curtail emissions and reduce environmental impacts. Conventionally, carbon calculators have been developed and employed to assess and report on emissions as well as the cost implications associated with the activities encompassed by roadworks and streetworks. However, the emissions and other impacts associated with different design options for roadworks and streetworks extend beyond the boundaries of their immediate activities (the 'systems of interest'). Roadworks and streetworks are interconnected with various components of urban systems and other infrastructure systems. This interconnectedness spans across environmental, societal, and economic systems, introducing an array of interdependencies across different scales such as posing a threat to the street scene (Goodwin, 2005), increasing road congestion (City of London Corporation, 2011) and increasing the cost of congestion to the region and the country (Haylen & Butcher, 2019), leading to substantial increases in social, economic, and environmental costs (Parker, 2008; City of London Corporation, 2011). Current carbon calculators, albeit valuable, are insufficient to encompass the broader system-level aspects that these activities affect. Here we present research into developing an All Consequences Calculator that includes the broader emissions and prioritizes avoiding environmental harm and negative socio-economic consequences.

Keywords All consequences calculator · Roadworks and streetworks · Design options · Sustainability · Net zero

K. Jutipanya (✉) · C. D. F. Rogers · E. J. S. Ferranti
University of Birmingham, Birmingham, England
e-mail: kxj234@student.bham.ac.uk

C. D. F. Rogers
e-mail: c.d.f.rogers@bham.ac.uk

E. J. S. Ferranti
e-mail: e.ferranti@bham.ac.uk

© The Author(s), under exclusive license to Springer Nature Switzerland AG 2026

J. Dsilva (ed.), *Redefining Sustainability*, https://doi.org/10.1007/978-3-031-88376-7_19

Introduction

Roadworks and streetworks contributed 9% to the UK's 2022 carbon emissions, equivalent to 75,000 flights from London to New York (Local Council Roads Innovation Group, 2023). Despite design options having diverse impacts, conventional cost–benefit analyses prioritize less expensive options, often overlooking broader aspects. These activities extend beyond immediate impacts, significantly affecting road ecosystems. For instance, trenching operations disrupt road access, impacting economic aspects of the local area and causing environmental damage. Such impacts are frequently neglected by carbon calculators and feasibility studies.

Choosing a design option based on comprehensive impact concepts necessitates a transformative approach emphasizing functionality, sustainability, resilience, and liveability. This shift requires exploring new business models to avoid unintended adverse consequences. Achieving Net Zero requires a methodology rooted in engineering design values and considering stakeholder aspirations, especially road users. This approach transforms engineering practices, orienting them towards sustainability, and benefits diverse individuals, challenging conventional feasibility studies. This inspired the aim of this study: to develop a comprehensive impact calculator for roadworks and streetworks that acknowledges the interdependencies between systems and their impacts, thereby maximizing their potential and ensuring the long-term sustainability and resilience on a context-dependent basis.

Integrating holistic elements in early planning underscores the significance of social, economic, and environmental factors, empowering engineers to shape more sustainable cities. This model aligns with the United Nations Sustainable Development Goals, promoting equity in economics and corporate realms (United Nations, 2015).

Carbon Assessment Tools Anatomy

The review of carbon assessment tools for roadworks and streetworks has identified over 15 well-established tools tailored for carbon assessment within the domain of infrastructures. These tools have been meticulously crafted by esteemed organizations, Highways in the UK, the European Federation of Foundation Contractors, and the UK Transport Research Laboratory (representing Highways England and Eurobitume), among others (Hussain et al. 2016; Moncaster & Symons, 2013). While the tools cover life cycle stages, including refurbishment, they lack focus on end-of-life benefits like reuse (Highways, 2021). For Net Zero emissions, it is crucial to address beyond in-use emissions.

Consequences Assessment

More than two decades of research on roadworks and streetworks lacks a systematic approach to understanding their consequences on urban systems. Notably, a system for managing urban infrastructure assets has been proposed, integrating modular ontology and rule-based modelling with qualitative likelihoods assigned by experts to address knowledge uncertainties (Wei et al., 2020). The outcomes, represented visually in an impact matrix table, categorize significant societal consequences into low, medium, serious, and high impact levels.

Streetworks' impact on urban systems, despite sporadic attention, involves utility works limiting access, potentially endangering pedestrians, including disabled individuals and children, and incurring costs through street tree damage (Marvin & Slater, 1997). In sewer management, failures disrupt utilities, causing traffic congestion and supply disruptions (Barber, 1997). An urban planning case study recommends policy reviews for coordinated street management, and while it details design-to-governance interconnections, it lacks specific consequences on the local area (Addison & Webber, 1996).

Area of Impacts

- Accidental impacts

Each primary user is susceptible to unintended consequences arising from roadworks and streetworks. Numerous scholarly investigations and legal bodies (Addison & Webber, 1996) have initiated discussions regarding the hazards posed by construction equipment and vehicles in road-related work zones, including reduced visibility and modified speed limits, which can lead to accidents among road users (Boesen Law, n.d.; Isern, 2021).

- Other impacts

The system's interconnected nature impacts all street users, particularly road users, pedestrians, and cyclists, experiencing travel delays due to construction activities with significant economic and social repercussions (Goodwin, 2005). Streetworks scrutiny encompasses construction-related costs, including operational direct and indirect expenses, and documented cases of tree and environmental damage (Hayes et al., 2012). Road degradation causes vehicle damage, escalating costs for road users, and heightened potential for implementing access charges, increasing the risk of service disruptions. These factors can adversely affect local businesses' sales (Burtwell et al., 2006).

The methodology entails meticulous stakeholder identification within the planning and construction processes of streetworks and roadworks, herein considered as individual participants. This involves also considering individuals who utilize the streets as users, thereby providing insights into the complexities and their aspirations regarding the streets. Through direct engagement, aspirations are extracted from semi-structured interviews and subjected to thematic analysis to discern key sustainable themes. A tailored assessment method discerns tangible and intangible impacts, using system mapping to illustrate interconnected systems affected by various design options for roadworks and streetworks in the specific context of the area where the works will be carried out. The 'All Consequences Calculator' aligns with the United Nations Sustainable Development Goals, providing insights into costs and overall value for iterative engineering processes.

Methodology

Participant and User Identification

This initial step entails a comprehensive identification of all participants, which in this study consists of the professional stakeholders and road users intricately interwoven within the complex system of roadworks and streetworks. In-depth recognition of these stakeholders is fundamental to the subsequent phases of the analysis, enabling the delineation of the analysis area and furnishing recommendations regarding focal areas for enhancement or areas susceptible to alterations and substantial impact.

Aspirations and Key Themes Identification

In this phase, stakeholders and users are directly engaged to identify key aspirations and fundamental themes in the urban system, using thematic analysis. These primary influencers, integral to the social, economic, and environmental dimensions, express diverse expectations for future street design, reflecting varied needs. Achieving sustainable design requires significant changes in technical and governance approaches. This holistic method optimizes street corridor systems, enhancing efficiency and resilience to short-term and long-term changes. The analysis categorizes themes into society, economy, and environment, encompassing safety, health, wellbeing, environmental considerations, land use planning, inclusive and resilient design, transportation, accessibility, technology, advocacy, and governance—core elements offering innovative approaches for individual and societal benefits, prioritizing primary stakeholders.

Assessment Method Development

A tailored assessment method is being developed to distinguish direct, tangible impacts from indirect, intangible impacts across specific activities. It is crucial to formulate this method while accommodating constraints imposed by contractual agreements and yet in a constraint-free context to ensure a holistic evaluation of impacts. Themes and specific focus areas in Table 19.1 align with aspirations articulated by workshop participants, including users and stakeholders, reflecting the value placed on streets. This exercise yielded discrete expressions of value clustered under themes and focus areas.

In a separate system mapping exercise, interdependencies between four streetworks design options (Trenching, Minimum Dig Technologies, Trenchless Technologies, and Advanced Robotics) were identified. The Trenching system map of consequences (Fig. 19.1) illustrates the impact of the trenching design option on interconnected systems, serving as the initial point of contact for those likely to be influenced by trenching activities. Each point of interdependence holds inherent value that can be either diminished or enhanced by the trenching process.

Elements in Table 19.2 represent potential impacts in an 'all consequences' calculation for Trenching. After an initial calculation, sustainability changes can be included in alternative designs, modifying consequences based on contextual nuances.

Table 19.1 Summary of themes

No.	Themes arising from the workshop[a]		
	I. Common themes	II. Appear to be priority	III. Underlying importance
1	Design standards	Land use strategy	Environment
2	Transportation and accessibility	Transportation and accessibility	Technology
3	Technology	Governance	Governance
4	Land use strategy		Health and wellbeing
5	Other considerations		

[a] The Road to Net Zero workshop, organized by the Road to Net Zero research team, took place on April 20, 2023, at the University of Birmingham. This initiative, the Road to Net Zero project, is financially supported through Lane Rental Charges, with the aim of reinvesting in ways that have a positive impact on both society and the environment. This collaborative endeavour involves Transport for London, the Highway Authorities and Utilities Committee (UK), the University of Birmingham, and EA Technology

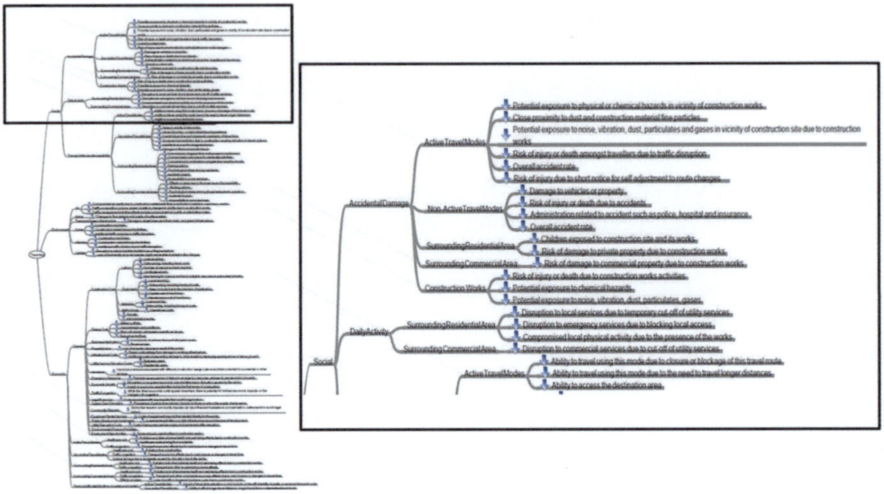

Fig. 19.1 Example of the trenching system map of consequences (the detail is not intended to be read)

All Consequences Calculator

In the ultimate phase, the focus shifts to value generation in Social, Economic, and Environmental domains, and alignment with the United Nations SDGs. A thorough assessment across these dimensions ensures a nuanced evaluation of roadworks and streetworks. The crafted system map offers investors a holistic view of costs, consequences, and overall value linked to the chosen design option, like trenching. Subsequent engineering iterations can then be used to refine processes and materials, amplifying positive outcomes while reducing or eliminating negative impacts.

Conclusions

To achieve Net Zero, a transformative shift from carbon-intensive methods is essential, starting at the stage of project planning. Relying solely on carbon cost assessments for current practices is inadequate, given the complex interconnections in the urban landscape—streets, trees, households, transportation, underground assets—shaping daily life. Viewing these elements as part of the urban metabolism highlights the need for comprehensive planning. Consideration of carbon costs, from initial planning to completion, for roadworks and streetworks' design options is crucial. However, the intricate connections demand thorough examination to progress towards Net Zero in the most effective manner possible. Initiating this transformation in project planning acknowledges the interconnected urban systems and contributes

Table 19.2 Elements with potential impacts on all consequences calculations for trenching

No.	Element[a]
1	Personal affordability of the consequences of the works
2	Surrounding area—commercial impacts
3	Surrounding area—residential impacts
4	Non-active transport mode impacts
5	Active transport mode impacts
6	Business disruption
7	Traffic congestion
8	Health and wellbeing impacts
9	Employment opportunities
10	Environmental fines
11	Utility relocation costs
12	Public infrastructure investment
13	Equipment rental
14	Community disputes
15	Operator supply chain and manufacturing impacts
16	Legal expenses
17	Economic growth
18	Emergency response
19	Insurance costs
20	Utility disruption
21	Habitat fragmentation
22	Biodiversity loss
23	Noise
24	Vibration
25	Tree/vegetation impacts
26	Water quality
27	Air quality
28	Waste
29	Accessibility to local facilities, services, or transportation
30	Temporary loss of parking
31	Travel time
32	Local activity disruption
33	Physical activity disruption
34	Accidents on site, accidents indirectly caused by the works or accidents otherwise arising from the works
35	Exposure to hazards

[a] The information contained within table is an integral component of The Road to Net Zero project. This initiative is financially backed by Lane Rental Charges and represents a collaborative effort among several key stakeholders, including Transport for London, the Highway Authorities and Utilities Committee (UK), the University of Birmingham, and EA Technology

to a sustainable future. The key concepts underpinning the development and use of an All Consequences Calculator are:

Shift Towards Net Zero: Emphasizes the necessity of transforming approaches and strategies from conventional, carbon-intensive methods to achieve Net Zero emissions.

Visualizing the Urban Metabolism: Introduces the concept of urban metabolism, where elements flow through and impact the city, underlining the need for comprehensive planning.

Interconnected Systems: Highlights the ongoing discourse about the interdependencies of urban systems, urging for a more systematic approach to represent the consequences of engineering interventions in complex systems.

Benefits of Innovative Engineering: Discusses the advantages of less intrusive engineering options, such as trenchless or robotic designs for streetworks, in reducing traffic delays, improving accessibility, and contributing to better air quality.

Systematic Research Methodology: Introduces a research methodology that serves as a guide to navigate towards Net Zero, ensuring a holistic understanding and representation of consequences and benefits.

Context-Dependent Analysis: Acknowledges the context-dependent nature of the analysis, considering specific issues related to locations, times, transport corridors, and design options.

Balancing Value: Strives for a balance of value in influencing new business models, promoting sustainable and user-centric street designs, addressing social, economic, and environmental considerations.

Implementation of Insights: Encourages the implementation of insights to forge streets aligned with community aspirations, fostering resilience, safety, and environmental sustainability on the path to Net Zero.

This comprehensive and systematic approach hinges on the contextual setting itself. It strives to strike a balance of value in a manner that profoundly influences new business models, substantiated by discernible evidence. The concise summaries presented above encapsulate a systematic pathway towards the realization of sustainable and user-centric street designs. This approach places a strong emphasis on addressing social, economic, and environmental considerations, leveraging technological advancements, and fostering collaborative practices. By implementing these invaluable insights, we can effectively forge streets that seamlessly align with the aspirations of the community, thus fostering resilience, safety, and environmental sustainability within the urban fabric, all while advancing towards Net Zero.

References

Addison, L., & Webber, S. (1996). Beyond the boulevards—the reality of Hounslow an analysis of the role of local authorities in seeking to improve environmental quality. *Landscape and Urban Planning, 35*(2–3), 123–135.

Barber, C. (1997). Chapter 6. Traffic management and public relations. In *Sewers rehabilitation and new construction repair and renovation*. Arnold; Wiley.

Boesen Law. (n.d.). *Increase in construction zone car accidents*. Boesen Law. Retrieved October 29, 2023, from https://www.boesenlaw.com/blog/2022/september/increase-in-construction-zone-car-accidents/

Burtwell, M. H., Evans, M., McMahon, W., & Kingdom, U. (2006). *Minimising street works disruption: The real costs of street works to the utility industry and society*. UK Water Industry Research Limited, London, UK.

City of London Corporation (2011). *Out of the jam: Reducing congestion on our roads: ninth report of session 2010–12*. The Stationery Office. https://publications.parliament.uk/pa/cm201012/cmselect/cmtran/872/872vw20.htm

Goodwin, P. (2005). Utilities' street works and the cost of traffic congestion. https://www.uwe.ac.uk/research/centres-and-groups/cts/research-themes/the-travel-environment/street-works-and-congestion. Accessed February 10, 2024.

Hayes, R., Metje, N., Chapman, D. N., & Rogers, C. D. F. (2012). Sustainability assessment of UK streetworks. *Proceedings of the Institution of Civil Engineers—Municipal Engineer, 165*(4), 193–204.

Haylen, A., & Butcher, L. (2019). *Street works in England*. UK Parliament. House of Commons 2011 Roads. https://commonslibrary.parliament.uk/research-briefings/cbp-8500/

Hussain, R. S., Enoch, M., Ruikar, K., Brien, N., & Gartside, D. (2016). Street work policy in England, UK: Insights from stakeholders. *Infrastructure Asset Management, 3*(2), 61–70.

Isern, K. (2021, March 11). *How road construction leads to more car accidents*. Amarillo Lawyers.

Local Council Roads Innovation Group. (2023, June 16). *Road to Net Zero: Project update*. LCRIG. https://lcrig.org.uk/news/road-to-net-zero-project-update/

Lovell, Lovell, Isern & Farabough, LLP. https://www.lovell-law.net/blog/personal-injury/how-road-construction-can-lead-to-an-increase-in-car-accidents/

Marvin, S., & Slater, S. (1997). *Urban infrastructure: The contemporary conflict between roads and utilities*. Pergamon.

Moncaster, A. M., & Symons, K. E. (2013). A method and tool for "cradle to grave" embodied carbon and energy impacts of UK buildings in compliance with the new TC350 standards. *Energy and Buildings, 66*, 514–523.

National Highways. (2021). *Carbon emissions calculation tool—National Highways*. Nationalhighways.co.uk. https://nationalhighways.co.uk/suppliers/design-standards-and-specifications/carbon-emissions-calculation-tool/

Parker, J. (2008). Briefing: The real cost of street works. *Proceedings of the Institution of Civil Engineers—Transport, 161*(4), 175–176.

United Nations. (2015, September 25). *Transforming our world: The 2030 agenda for sustainable development*. United Nations. https://sdgs.un.org/2030agenda

Wei, L., Du, H., Mahesar, Q., Al Ammari, K., Magee, D. R., Clarke, B., Dimitrova, V., Gunn, D., Entwisle, D., Reeves, H., & Cohn, A. G. (2020). A decision support system for urban infrastructure inter-asset management employing domain ontologies and qualitative uncertainty-based reasoning. *Expert Systems with Applications, 158*, Article 113461.

Chapter 20
Evaluating CO_2 Emissions of an Industrial Warehouse: A Comparative Analysis Between Autodesk Green Building Studio and Autodesk Tally

Jhanvi Sunil Sabhnani, Syed Ibrahim Ali, and Vidya Mohanan

Abstract In our contemporary era, heightened concerns regarding climate change have underscored the critical role of sustainability within the construction industry. Building Information Modeling (BIM) has emerged as a potent instrument for environmental analysis, offering substantial promise in this regard. The fusion of BIM with Life Cycle Assessment (LCA) facilitates automated evaluations of environmental impacts, especially concerning energy consumption in material production, transportation, and construction processes. This amalgamation enables the determination of embodied energy and carbon emissions of construction materials, enabling designers to optimize energy usage and diminish the environmental impact of buildings right from the initial design phase. The evaluation of materials' environmental impact holds immense significance in achieving energy efficiency and elevating sustainability within the construction sector. Consequently, this study advocates for a user-centric approach that integrates LCA and BIM to appraise and mitigate the embodied energy in a prototype design for an industrial warehouse situated in Dubai UAE. To advance sustainable design solutions, simulations were conducted utilizing Autodesk Revit, a BIM authoring tool, in conjunction with LCA analysis. Additionally, a comparative analysis of embodied energy and carbon emissions was performed using Autodesk Green Building Studio and Tally (Revit Plugin) to identify optimal sustainable design strategies.

Keywords Sustainability · Embodied energy · Autodesk Revit · Autodesk green building studio · Autodesk Tally

J. S. Sabhnani (✉) · S. I. Ali · V. Mohanan
Amity University Dubai, Dubai, United Arab Emirates
e-mail: jhanviS1@amitydubai.ae

S. I. Ali
e-mail: syedibrahimA@amitydubai.ae

V. Mohanan
e-mail: vmohan@amityuniversity.ae

© The Author(s), under exclusive license to Springer Nature Switzerland AG 2026
J. Dsilva (ed.), *Redefining Sustainability*, https://doi.org/10.1007/978-3-031-88376-7_20

Introduction

The early stages of design are crucial in minimizing a building's environmental impact. Architects and construction professionals can prioritize these stages to make informed decisions that promote sustainable development. Efforts have been made to combine BIM software with LCA methodology and environmental optimization techniques to facilitate environmentally conscious decision-making and integrate sustainable practices into the design process. Within the built environment, sustainability pertains to the construction and maintenance of structures to reduce negative environmental effects and optimize social and financial gains. Sustainable building practices aim to minimize pollution, conserve energy and water, and minimize waste using materials and construction techniques. As the benefits of Building Information Modeling (BIM) become widely recognized by building design professionals, they will assume a more prominent role in enhancing the environmental performance of buildings throughout their entire life cycle. Utilizing information like building orientation, and thermal mass, BIM can assist in identifying locations that can be optimized for lower energy use.

The UN predicts that there will be 9.7 billion people on the planet by 2050 (AutoDesk). To meet global demand and contribute to the creation of smarter, more resilient places, the global architecture, engineering, and construction sector must look to smarter, more efficient ways to design and build. Design and construction teams can operate more productively thanks to building information modeling software, and they can also use the data they generate to support operations and maintenance tasks. For this reason, Building Information Modeling requirements are growing globally.

Therefore, the objectives of this research paper were to design a prototype of an industrial warehouse on Autodesk Revit 2024 and to study the comparative analysis between Autodesk Green Building Studio and Autodesk Tally, henceforth, to study potential strategies to reduce the structure's energy consumption and carbon footprint.

Literature Review

Human activities, such as burning fossil fuels and deforestation, emit harmful pollutants impacting the environment (Buchanan, 1994). The construction industry, a major pollutant and energy consumer, releases sulfur dioxide and fine particulates during fossil fuel power generation and material degradation (Zimmermann, 2014). Construction's environmental impact stems from various sources, including individual projects, raw material demand, processing, and completed projects (Anigbogu, 2005), affecting ecosystems, atmosphere, plant life, soil, and climate.

Manufacturing construction materials contribute to pollution, with notable impacts like GHG emissions, dust, gas emissions, noise, waste generation, water consumption, and air pollution (Chuai, 2021). Mitigating these issues requires sustainable practices, such as resource conservation, waste reduction, energy efficiency, and eco-friendly materials.

Buildings have embodied and operational energy consumption (Han, 2013), influenced by factors like building type, standard development, climate region, and technology (Ajayi, 2019). Residential and commercial buildings in the US consume significant energy, highlighting the need for sustainable construction methods (Azari, 2019). Research emphasizes that conventional methods have higher embodied energy and carbon footprints than non-conventional ones, advocating strategies like local sourcing, energy-efficient designs, and natural ventilation (Lu, 2017; Uddin, 2021). The construction industry aims to shift toward low-energy materials, energy-efficient designs, and renewable energy for more sustainable buildings (Zhang, 2018). Saving energy during a building's lifecycle drastically reduces life cycle costs, emphasizing the importance of sustainable practices and renewable materials in Life Cycle Assessment (Najjar, 2017).

Collaboration among project stakeholders during construction faces limitations with traditional methods like 2D CAD designs (Bughio, 2021). Advanced technologies, including 3D and 4D CAD applications and BIM tools, enhance visualization, scheduling, and project understanding (Iqbal, 2022). BEM, during a building's lifecycle, improves energy efficiency, offering crucial insights gained during the planning phase (Maglad, 2023). While 3D CAD has limitations, BIM's integrated and parametric approach ensures consistency across views, promoting collaboration and problem-solving throughout the project lifecycle (Abbondati, 2020; Azhar, 2015). Building Information Management, as per Krygel and Nies (Krygiel, 2008), enhances design through building massing, energy modeling, and building orientation (Ghedamsi, 2014; Moakher, 2012).

Green BIM supports sustainable building development, emphasizing the integration of design information and collaboration (Wong, 2015). The construction industry, with its socio-economic importance, must balance global and local construction materials. Green products, made from eco-friendly materials, aim to minimize environmental impact (Ramesh, 2010). Despite a potential increase in embodied energy, passive and active technologies reduce a building's life cycle energy demand, contributing to global efforts to reduce energy usage (Weymouth, 2018).

A case study (Al-Ghamdi, 2016) was conducted at Magee-Women's Hospital, where three software tools, namely Tally, SimaPro, and Athena, were utilized to assess the material and energy consumption over the building's projected 60-year life cycle, as per the LEED minimum requirement. The data obtained from these software tools revealed that the energy consumption exceeded the LEED requirements by approximately 10%. This outcome is notable considering that hospitals have high energy consumption within the building sector, with the highest energy usage per unit area compared to other building types. Factors, such as extensive heating, cooling, round-the-clock operations, and the utilization of numerous medical equipment, contribute to the elevated energy intensity and carbon dioxide emissions in hospitals, surpassing commercial office buildings by more than 2.5 times annually, resulting in over 146 kg/m^2 of CO_2 emissions (Al-Ghamdi, 2016; Balaras, 2007). In terms of material composition, concrete and masonry make up around 65% of the hospital's total mass. However, the most significant environmental impacts are attributed to metals and finishes due to their embodied energy. The embodied energy, as opposed to operational energy, played a substantial role in the hospital's CO_2 emissions, accounting for 99% of the total global warming potential.

Amidst increasing energy costs and environmental concerns, there is a growing demand for resource-efficient and ecologically sound construction practices (Santos, 2018). However, tools like Tally have limitations in material selection and offer generic information, hindering accurate LCA analysis (Santos, 2018). Tally's assessment based on Tracy 2.1, specific to the US, further restricts its global applicability.

Methodology

The methodology adopted in this research paper is carried out in three stages—modeling and energy analysis on Revit 2024, analyzing the structure on Autodesk Green Building Studio, and analyzing the structure on Tally software.

Autodesk Revit 2024

Autodesk Revit is a widely employed 3D building information modeling program in the architectural, engineering, and construction sectors. This feature-rich tool integrates plumbing, electrical, and civil plans into a single drawing, allowing for early detection and correction of conflicts or errors during the design phase. Through its building elements like walls, roofs, windows, and floors, Revit generates comprehensive 3D models, including conceptual massing capabilities for early building form and orientation modeling. Beyond architectural features, Revit extends its capabilities to MEP and structural design, making it a crucial tool in the modeling phase for energy simulation within the AEC sector.

In this research paper, an industrial warehouse is modeled on Revit 2024. The general details of the model are shown in Table 20.1.

Figure 20.1 represents the column grid spacing in mm, and Fig. 20.2 represents the elevation view of the warehouse.

The 3D view of the warehouse is shown in Fig. 20.3; upon completing the model on Revit, the energy analysis was conducted on the model. The energy analysis was carried out on Autodesk Revit by creating spaces, thus creating a space schedule in which the area and volume of the model are displayed as 503 m^2 and 1225.97 m^3, respectively, and the condition type is chosen as "naturally vented only". For performing the energy analysis, the energy setting must be modified. Therefore, on the analyze tab, select energy settings. Modifying the mode to "use conceptual masses and building elements", then clicking "edit" under other options. A pop-up screen appears with the advanced energy settings; at this stage, the building type is

Table 20.1 General details of model

S. No.	Details	Dimensions
1	Size of warehouse	20 m × 24 m
2	Elevation of warehouse	5200 mm
3	Height of truss	1200 mm
4	Roof pitch angle	13.5°
5	Clear span	10 m
6	Number of bays	2
7	Bay spacing	6 m
8	Location	Al Quoz industrial area, U.A.E (chosen test location)

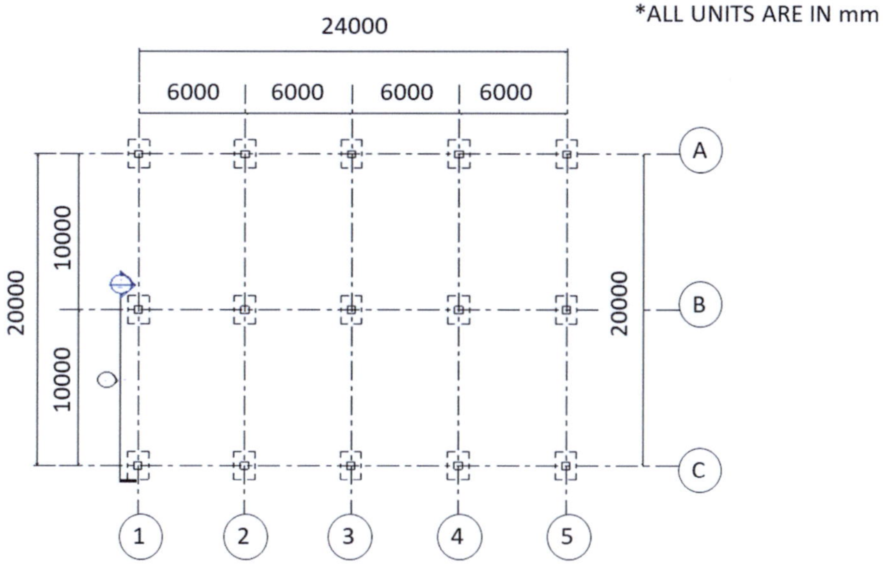

Fig. 20.1 Column grid spacing of the warehouse in mm

chosen as "warehouse", and the building operating schedule is set to "12/6 facility" which refers to the building operating from Monday to Saturday but is closed at night and Sundays, and then click ok. In order to generate the energy model, on the analyze tab, under the energy optimization panel click on generate energy model. A pop-up window appears, and "use existing energy model" is chosen. Figure 20.4 represents the energy model created in 3D view.

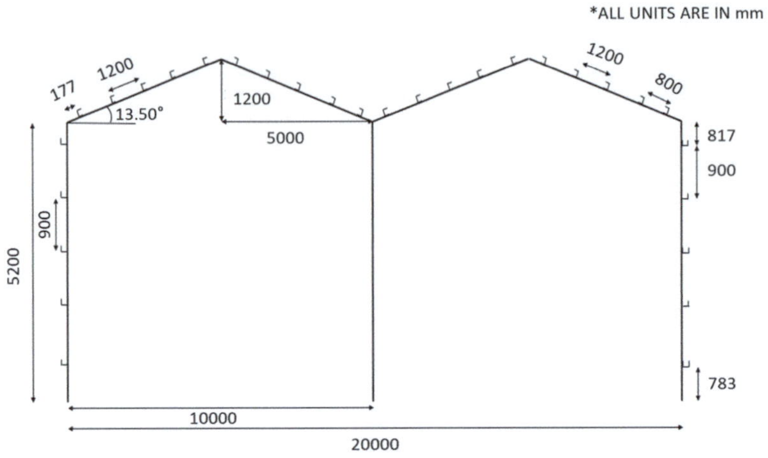

Fig. 20.2 Elevation view of the warehouse

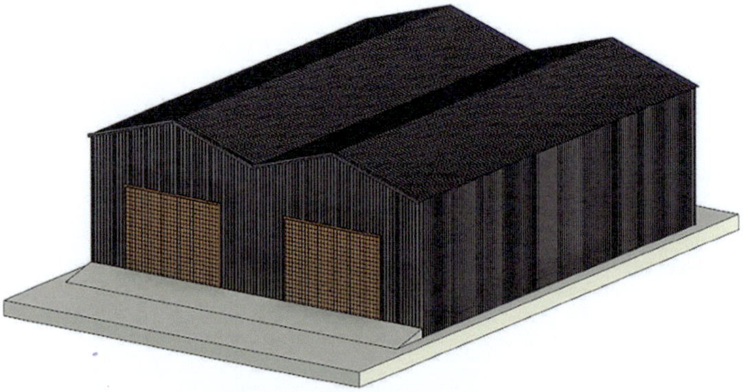

Fig. 20.3 3D view of the warehouse

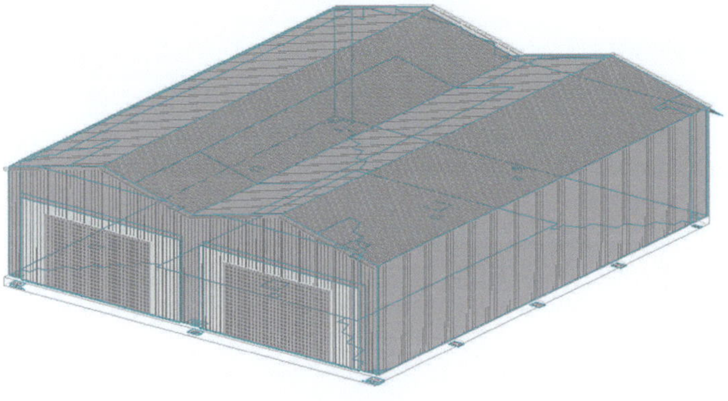

Fig. 20.4 Energy model created on Autodesk Revit in 3D view

Autodesk Green Building Studio

In a whole building energy analysis, all energy-consuming systems are considered, providing accurate energy cost estimates for financial and design decisions. Autodesk Green Building Studio, a cloud-based service, streamlines the process by incorporating defaults and assumptions based on building location and type. This versatile tool allows for the simulation of building performance, optimizing energy efficiency, and working toward carbon neutrality in the early design phase. Green Building Studio offers efficient and cost-effective design of high-performance structures compared to traditional methods. The service significantly reduces the time required for a comprehensive building energy study, providing clear guidance on impactful changes, and automatically assessing various architectural elements. Utilizing the cloud service, Green Building Studio leverages the DOE-2 simulation engine for reliable results in energy use, water use, and carbon emissions analysis.

To find the environmental impact of the structure on Autodesk Green Building Studio, the following steps were adopted. Firstly, the location of the project was set to Al Quoz Industrial Area, U.A.E, and the closest weather station was chosen. Secondly, the Revit file was exported to gbXML file. Then, the creation of a new project was carried out on Autodesk Green Building Studio, wherein the type of building was defined as "warehouse", schedule as "12/6 faculty", and project type as "test project", and the location of the project was defined.

Setting up the project defaults was extremely crucial as it helps the user to choose the materials adequately and also helps to define utility rates. Therefore, the electric cost is input as AED 0.29/kWh and the fuel cost as AED 0.93/Therm. On completing this, the gbXML file was imported into the project.

Autodesk Tally

Tally, an Autodesk Revit plug-in, enables engineers and architects to assess the environmental impact of building materials for both comparative design assessments and full building analyses. Users can associate construction materials from the Tally database with building elements in a Revit model. Tally's methodology aligns with LCA standards, categorizing environmental assessments into product, construction, use, end of life, and module D stages, further broken down into sub-categories for detailed insights. Conducting a Life Cycle Assessment is essential for determining the embedded environmental impact of construction materials. While BIM theoretically provides a comprehensive bill of materials for assessment during project completion, real-world challenges in accurately representing material volumes highlight the need for precise environmental impact assessment tools in the design stage. Tally assists architects and engineers in exploring alternative designs early in the construction process, aiming to reduce the carbon footprint and utility bills.

The procedure adopted for Autodesk Tally is carried out by defining materials take-offs. Most materials in Tally are generic. Each material must be defined, assuming a suitable design life. The next step involves inputting the project details which include the project details, location, gross building area, and expected building life. Lastly, the report is saved and henceforth a detailed report regarding the environmental impacts of the project's development at every phase of the structure's existence.

Findings

Upon conducting the methodology, the data for carbon dioxide emissions was collected from Autodesk Green Building Studio and Autodesk Tally.

From the data collected on Autodesk Green Building Studio for the annual CO_2 emission, a bar graph is drawn as shown in Fig. 20.5. It can be understood that on an annual basis, the CO_2 emissions generated by the structure are 25.6 Mg. If design alternatives are added to the structure and renewable materials and sources are utilized in the structure, the CO_2 emissions can be reduced; hence, this is indicated by the negative sign used, i.e., the onsite renewable potential of – 33.9 Mg.

Tally established defined stages for the life cycle of a project. These stages include the manufacturing stage (A1-A3), construction stage (A4-A5), building operation stage (B6 and B7), end-of-life stage (C1-C4), and module D (Stine, n.d.).

The data obtained from Autodesk Tally is shown below in the form of a pie charts in Figs. 20.6 and 20.7.

As shown in Fig. 20.6, during the life cycle of a structure, the product stage (A1-A3) accounts for 64% of the total carbon emissions, making it the highest contributor. This stage involves the manufacturing and extraction of materials. The transportation

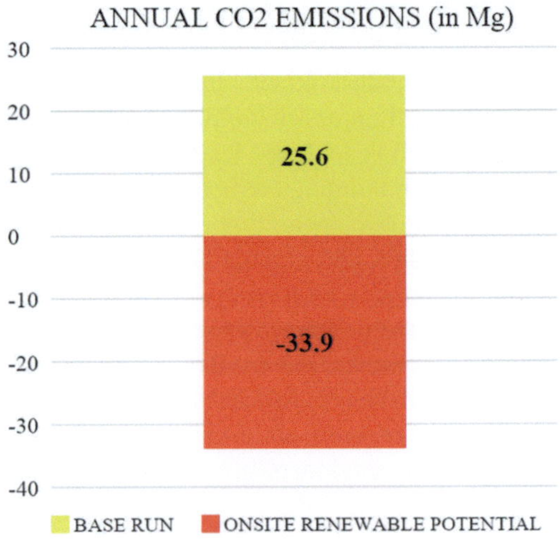

Fig. 20.5 Bar graph representing annual CO_2 emissions (in Mg) as obtained by Autodesk GBS

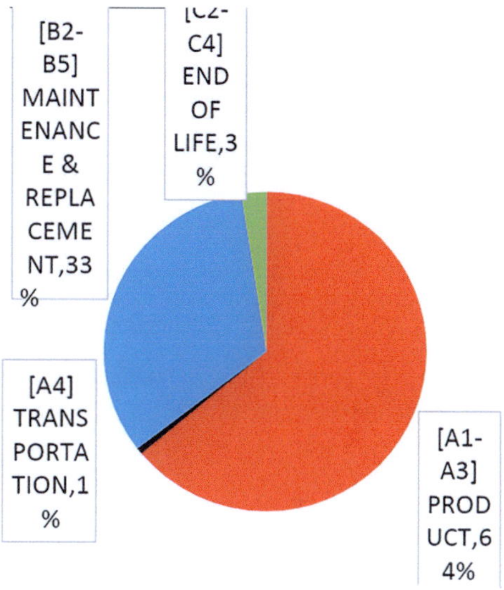

Fig. 20.6 Percentage of global warming emissions generated at each stage in the life cycle of the structure on Autodesk Tally

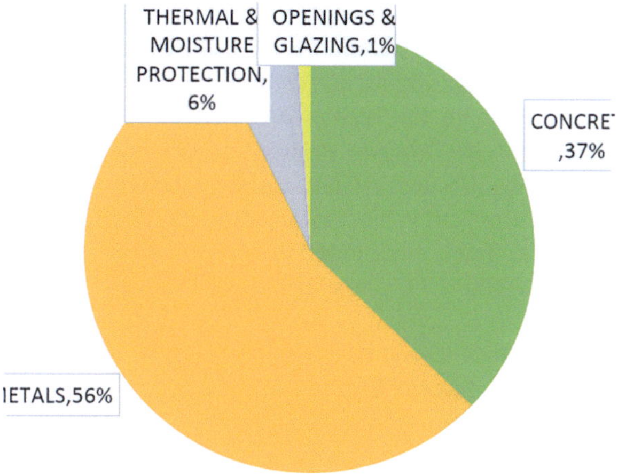

Fig. 20.7 Percentage of global warming emissions per division on Autodesk Tally

stage (A4) contributes only 1% of emissions and involves transporting materials to the construction site. The maintenance and replacement stage (B2-B5) make up 33% of emissions and consider the service life of materials. Lastly, the end-of-life stage (C2-C4) represents 2% of emissions and covers transportation to treatment facilities and the potential for recycling and disposal of materials.

As shown in Fig. 20.7, the highest carbon dioxide emissions, accounting for 56%, come from the metals used, including steel sections and metal sheets. As the structure is predominantly steel, it contributes significantly to the carbon footprint. Concrete contributes 37% of emissions, primarily from the high-grade concrete used for the foundation due to the need for strength in a sandy environment like the U.A.E.

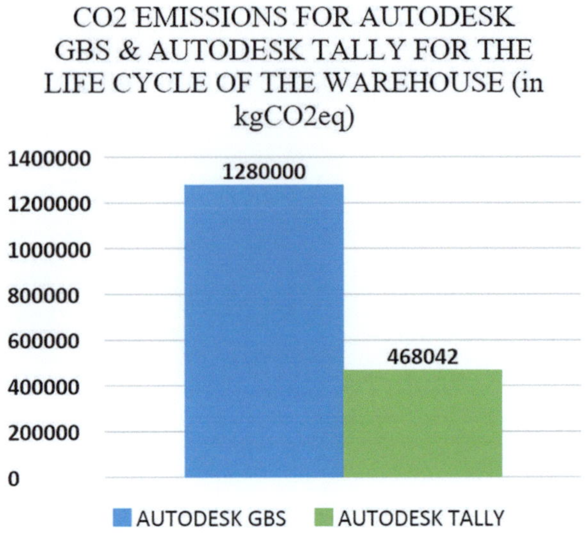

Fig. 20.8 Percentage of global warming emissions per division (in $kgCO_2$ eq)

Thermal and moisture protection, including coatings and DPC coatings, generates 6% of emissions. The remaining 1% is attributed to the openings and glazing, including the two shutter doors.

To compare the results obtained for CO_2 emissions between Autodesk GBS and Autodesk Tally, an assumption is made that the annual emissions obtained on Autodesk GBS are constant for the entire life cycle of the warehouse for 50 years.

Based on the values obtained above, for CO_2 emissions in Autodesk GBS and Autodesk Tally, a bar graph is drawn as shown in Fig. 20.8.

The higher CO_2 emissions from Autodesk GBS compared to Autodesk TALLY can be due to several factors. GBS calculates emissions based on on-site fuel use and regional electricity sources, whereas TALLY associates emissions with categorized environmental impacts. GBS utilizes CARMA data for non-US locations, while TALLY employs the TRACI 2.1 characterization scheme by the US EPA. Additionally, GBS lacks the option to input quantity take-off details for materials, unlike TALLY. Furthermore, GBS reports emissions annually, leading to a vague approximation, whereas TALLY provides a detailed breakdown of emissions for each stage of the structure's life cycle. These differences in data and features justify the observed variation in emission values between Autodesk GBS and Autodesk Tally.

Conclusion

The results obtained from performing the analysis helped to understand the carbon footprint of the structure. The carbon footprint of an industrial steel warehouse is largely influenced by the manufacturing and extraction stages, which account for 64%

of emissions. In the case of a steel warehouse, metals contribute to approximately 56% of the emissions. By recognizing these high-impact areas, designers and engineers can focus their efforts on finding sustainable alternatives or optimizing processes to reduce emissions.

To create energy-efficient buildings and reduce operational costs, a combination of active design strategies (e.g., solar panels, energy-efficient mechanical systems) and passive design techniques (e.g., optimal orientation, proper material selection) can be implemented. Additionally, reducing the carbon footprint of construction can be achieved by choosing locally available materials that require less energy and money for extraction and manufacturing, while considering recyclability and maintaining structural strength. Choosing materials that can be easily recycled or repurposed at the end of their lifespan is crucial in minimizing waste and reducing the environmental impact of construction projects. At the end of a structure's life, steel components can be easily and economically recycled without compromising their quality, enabling easier deconstruction, dismantling, and the salvage or reuse of steel elements in construction.

Difficulties arose in the study during the installation of exhaust fans in the modified model with shutter doors. The chosen software, Autodesk GBS, lacked international location data, leading to the use of United States settings for carbon footprint analysis. To mitigate the impact of default HVAC system values in Autodesk GBS, extreme heating and cooling set points were implemented. However, describing components in Autodesk Tally and conducting quantity take-offs proved time-consuming, requiring extensive research.

To address these challenges, an alternative software called TallyCAT can be utilized. TallyCAT is an Autodesk Revit plug-in that builds upon Autodesk Tally, offering more precise material take-offs, identifying high-carbon materials, and providing information on potential impact reduction and climate impact statistics. This alternative software streamlines the evaluation process and enhances data accuracy.

References

Abbondati, F. (2020). I-BIM for existing airport infrastructures. *Transportation Research Procedia, 45*, 596–603.

Ajayi, S. O. (2019). Changing significance of embodied energy: A comparative study of material specifications and building energy sources. *Journal of Building Engineering, 23*, 324–333.

Al-Ghamdi, S. G. (2016). Green building rating systems and whole-building life cycle assessment: Comparative study of the existing assessment tools. *Journal of Architectural Engineering, 23*(1).

Anigbogu, N. (2005). Impact of construction activities on the environment. *Towards a sustainable built environment. 2nd National Conference.*

AutoDesk. (n.d.). Retrieved from https://www.autodesk.com/solutions/aec/bim/benefits-of-bim

Autodesk Green Building Studio. (n.d.). (Autodesk) Retrieved from https://gbs.autodesk.com/GBS/?redirectUrl=%2FGBS%2FProject

Azari, R. (2019). Life cycle energy consumption of buildings; embodied + operational. In *Sustainable construction technologies* (pp. 123–144).

Azhar, S. (2015). Building information modeling (BIM): Now and beyond. *AJCEB Incorporating Special Issue on BIM, 12*(4).
Balaras, C. A. (2007). HVAC and indoor thermal conditions in hospital operating rooms. *Energy and Buildings, 39*(4), 454–470.
Buchanan, A. H. (1994). Energy and carbon dioxide implications of building construction. *Energy and Buildings, 20*(3), 205–217.
Bughio, M. (2021). Impact of passive energy efficiency measures on cooling energy demand in an architectural campus building in Karachi, Pakistan. *Sustainability, 13*(13).
ChooseTally. (n.d.). Retrieved from https://choosetally.com/
Chuai, X. (2021). China's construction industry-linked economy-resources-environment flow in international trade. *Journal of Cleaner Production, 278.*
Ghedamsi, R. (2014). Contribution on buildings design with low consumption of energy incorporated PCMs. *Energy Procedia, 50*, 322–332.
Han, M. (2013). Embodied energy consumption of building construction engineering: Case study in E-town, Beijing. *Energy and Buildings, 64*, 62–72.
Iqbal, A. (2022). An overview on need for automatic structural detailing of line elements in BIM tools. In *1st International conference on advances in civil and environmental.* Pakistan.
Koeppel, S. (2007). *Assessment of policy instruments for reducing greenhouse gas emissions from buildings.* Central European University.
Krygiel, E. (2008). *Green BIM: Successful sustainable design with building information modeling.* Wiley Publishing Inc.
Lu, Y. (2017). Building information modeling (BIM) for green buildings: A critical review and future directions. *Automation in Construction, 83*, 134–148.
Maglad, A. M. (2023). BIM-based energy analysis and optimization using insight 360 (case study). *Case Studies in Construction Materials, 18.*
Moakher, P. E. (2012). Building information modeling (BIM) and sustainability—using design technology in energy efficient modelling. *Journal of Mechanical and Civil Engineering, 1*(2), 10–21.
Najjar, M. (2017). Integration of BIM and LCA: Evaluating the environmental impacts of building materials at an early stage of designing a typical office building. *Journal of Building Engineering, 14*, 115–126.
Philipp Rode, R. B. (2011). Buildings: investing in energy and resource. In *Towards a green economy: Pathways to sustainable development and poverty eradication. United Nations Environment Programme* (pp. 331–373). London School of Economics and Political Science.
Ramesh, T. (2010). Life cycle energy analysis of buildings: An overview. *Energy and Buildings, 42*(10), 1592–1600.
Santos, R. (2018). A validation study of a semi-automatic BIM-LCA tool. In *2º Congresso Português de building information modelling.* Lisboa.
Stine, D. (n.d.). *Tally—life cycle assessment—by KieranTimberlake.* (BIM Chapters) Retrieved from https://bimchapters.blogspot.com/2018/04/tally-life-cycle-assessment-by.html
Uddin, M. N. (2021). Building information modeling (BIM) incorporated green building analysis: an application of local construction materials and sustainable practice in the built environment. *Journal of Building Pathology and Rehabilitation.*
Weymouth, R. (2018). Principles for integrating the implementation of the sustainable development goals in cities. *Urban Science* (77).
Wong, J. K. (2015). Enhancing environmental sustainability over building life cycles through green BIM: A review. *Automation in Construction, 57*, 156–165.
Zhang, C. (2018). BIM-based investigation of total energy consumption in delivering building. *Advanced Engineering Informatics, 38*, 370–380.
Zimmermann, M. (2014). An innovative architectural strategy for the integration of energy issues into the design process of a commercial center in Switzerland. *Journal of Building Construction and Planning Research, 2*(37), 1147–1157.

Chapter 21
Influence of Heritage Design Elements on Sustainable Interior Environments: A Case Study of Al Seef Dubai

Vrushali Mhatre

Abstract Heritage design elements over the years in the UAE have undergone an evolutionary process, drawing inspiration from local heritage, cultural wisdom, and accumulated knowledge. These traditional design features contribute to shaping spaces that promote a sense of belonging and local culture within the UAE's built environment. The aim of the study is to identify these heritage design elements and reassure their crucial role in interior architecture to ultimately enhance the sustainability of interior-built environments. This research paper critically examines a recent urban development in Dubai, with a specific focus on the Al Seef heritage area, characterized by historic architecture, and a contrasting space featuring contemporary structures. The investigation identifies elements of local heritage seamlessly integrated into contemporary planning and design. The research will offer a crisp overview of the design strategy, emphasizing solutions for temperature control, humidity regulation, ventilation, and maximizing natural light to achieve sustainability goals of a built environment. By employing a mixed qualitative case study research method, the findings will be systematically organized into visual tables, providing detailed descriptions evaluation summary. Through a case study analysis, this research paper imparts valuable insights gathered from local heritage design elements. It underscores how contemporary interior architecture can effectively enhance the sustainability of any interior-built environments in modern settings. The findings contribute to a deeper understanding of the harmonious integration of heritage and sustainability in interior design, offering lessons applicable to future architectural endeavors in similar contexts.

Keywords Heritage design · Sustainable · Interior environments · Built environment · Al Seef Dubai

V. Mhatre (✉)
Heriot Watt University Dubai, Dubai, United Arab Emirates
e-mail: V.Mhatre@hw.ac.uk

© The Author(s), under exclusive license to Springer Nature Switzerland AG 2026
J. Dsilva (ed.), *Redefining Sustainability*, https://doi.org/10.1007/978-3-031-88376-7_21

Introduction

The scope of this research encompasses the exploration and analysis of heritage design elements within the context of sustainable interior environments in Dubai. The study delves into the historical and cultural significance of traditional architecture in Dubai and aims to assess the feasibility and impact of integrating these heritage elements into contemporary interior design practices. The geographic focus is on Dubai, but the findings may have broader implications for other regions seeking to balance cultural heritage with sustainability in their built environments.

The vernacular design process evolves in combination with natural and cultural factors, resulting in the embeddedness of a wide variety of sustainable ecosystems in heritage-based building designs. These traditional sustainable design solutions serve as evidence of the interaction between the natural and anthropogenic environment (Adam, 2012; Hartel et al., 2014). The environment that this paper is studying is the (built environment), the term-built environment, refers to the human-made environment that provides the setting for human activity, ranging in scale. From buildings to cities and beyond, it has been defined as "the human-made space in which people live, work, and recreate on a day-to-day basis" (Rudofsky, 1987).

Hence, the built environment encompasses all places and surroundings shaped or modified by people to meet their needs for living, organization, and expression.

The concept of heritage means that there is an inherent worth that is being conferred upon a place, an object, an institution, or a particular collection of actions that are made up from its relation to some aspect of the past, as interpreted and immortalized by a single or a set of communities (Al Qasimi, 2019).

Theoretical Underpinning

Recently, builders and architects have been more forthcoming in their design proposal of consideration of both Arabic heritage and environmental concerns. These expressions have been seen by some as evidence that there has more harmonious development overseen by Dubai's rulers in the last 20 years (Rizvi, 2017). Efficient heat-resistant materials are increasingly used in construction, combined with traditional Arabic designs. At a basic level, protection is needed from high temperatures, high humidity, high solar radiation, dust storms, and desert winds. This is heavily reflected in the traditional and current use of courtyard houses and arish (houses built of palm tree leaves). This can be seen in the historic buildings for which Dubai has attracted worldwide attention. (Wikipedia contributors, 2024 February 20).

This research originates from the challenge that, despite the historical use of heritage elements, their sustainable advantages for the interior-built environment have not been sufficiently investigated. **Research Aim**: To identify heritage design elements that integrate daylight, improve indoor air quality and thermal comfort while using sustainable ecofriendly local material.

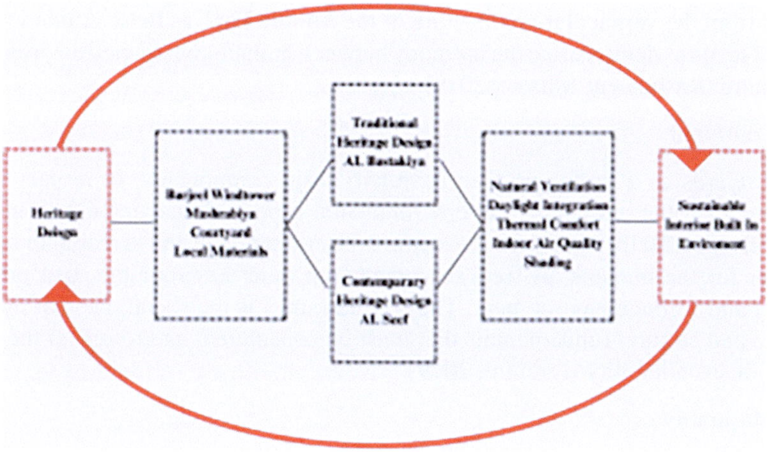

Fig. 21.1 Conceptual framework

Research Objective: 1. The objective of this study is to analyze the interplay between heritage design elements and sustainable practices in interior-built environment within the case study of Dubai Al Seef in context.

2.To combine all key heritage elements of the case study, like wind towers, mashrabiya, and courtyard to a built environment and spatial design to explore new paradigms of interior sustainability (Fig. 21.1).

The Al Seef area of Dubai is investigated in this case study research to understand the extent to which interior architecture design elements need to grow from the physical surroundings of the region interacting with the architectural patterns as well as the cultural setting around.

The present paper argues through an example of a case study that by integrating heritage design sustainable solutions into contemporary design can positively influence sustainability in today's-built environment. A literature review was conducted to explore the heritage design sustainable concepts and applications present in both heritage and contemporary designs, including books and journal articles. The commonly utilized traditional heritage elements in contemporary structures include Barajeel or wind towers, courtyards, and mashrabiya as well as traditional ornaments and decorative patterns, often recycled or imitated in contemporary design.

1.1 Barajeel

Due to an absence of mechanical cooling technologies, the Barajeel, used throughout the Gulf region, was a responsive design feature to circulate air through the house particularly during harsh summer months. Wind towers or catchers are tall rectangular structures with vertical openings on all sides, and the internal walls in these towers are diagonally arranged to ensure that breeze from outside is forced downwards, then up again to escape, thus creating a circulation of air through the rooms (Awad & Boudiaf, 2020). Contemporary designers have adapted modern wind towers from the vernacular architecture of the Middle East, as these dramatic architectural features demonstrate the harmony between human-constructed environments and nature (Rashdan & Mhatre, 2019).

1.2 Courtyard

Courtyards as a prototype plan have two basic components: to respect and to accommodate the climate as the environmental aspect of sustainability, since the major function for the courtyard is to control the environment and to maintain suitable weather for the humans, as well as to create an inner environment, that provides privacy and solitude environment. The second aspect is the social one that includes Islamic and cultural fundamentals that must be considered, and to reflect the social aspect of sustainability (Ibrahim, 2019).

1.3 Mashrabiya

Mashrabiya has two key features; one is sustainable and the other is social and cultural. The primary sustainable function of Mashrabiya serves as an important and sustainable device for efficient heating, cooling, and ventilation. This allows adjustments of the access of sunlight and protection from glare. The secondary sustainable functions might include sound insulation that may be generated by the wind or outside noises. Mashrabiya serves not only to cover openings and windows, but also to shield the occupants from the hard surrounding environment (Abdelkader & Park, 2018; Al Qasimi, 2019).

1.4 Traditional and Local Materials

Buildings incorporating wind towers along Dubai Creek were constructed of palm fronds—known as Bait Areesh—which was the standard material used to construct homes in the region pre-1900. The coral or sea stone homes of Dubai, locally known as Bait Morjan, were built using local materials, such as coral, shell, wooden joists, palm trunks, and woven mats (Ragette, 2003). In the 1950s, arish was a common facet of the urban landscape in the Gulf. Building houses from palm leaf was a community affair, with tribes and families working together to complete the residences (Chaves & Alexandra, 2020). Local materials characterized by a low-tech profile due to lack of production technology, therefore, have limited carbon footprint. Moreover, they are generated from local, natural resources so they do not emit any harmful gases—which enhances IAQ performance—and they are naturally biodegradable, which has a low impact on environmental pollution (Rashdan & Mhatre, 2019). Construction materials, whether they are made of palm or coral, are considered sustainable since they do not emit any harmful gases, enhance IAQ performance, and are biodegradable (Rashdan & Ashour, 2017).

Methodology

The research method is the analysis of heritage design elements and its application in the Al Seef heritage district of Dubai. In this paper, a descriptive analysis is used based on review of literature and field studies open observation and design details (refer Fig. 21.2). The research employs a grounded theory methodology which includes data collection from the case study and analysis. Hanson (1965) asserted that what is experientially true is already theoretically defined and is an important theoretical point of departure. Nothing is theoretically more stimulating than a well-functioning practice. Therefore, there is a need for theoretical contemplation on successful practices in the field of a sustainable interior-built environment (Yahya & Hassanpour, 2022). In the first stage of the study, the role of heritage design elements was analyzed. To achieve this, a two-stage strategy was adopted, and the first stage is to conduct a research study which includes a review and analysis of regional and local examples of the case study, which relied on strategies for sustainability concepts in the process of designing and creating them and applying the criteria extracted from the review of the literature. In the second phase, a methodical framework for sustainable strategies integrated is highlighted. Finally, the inclusivity, adaptability, and suggestiveness of the developed methodical framework are verified through the case of Al Seef in Dubai. The study also explored visual examples that were aimed at preserving the values, meanings, and flexibility of sustainable design elements that maintain a vernacular design identity while adapting to contemporary trends.

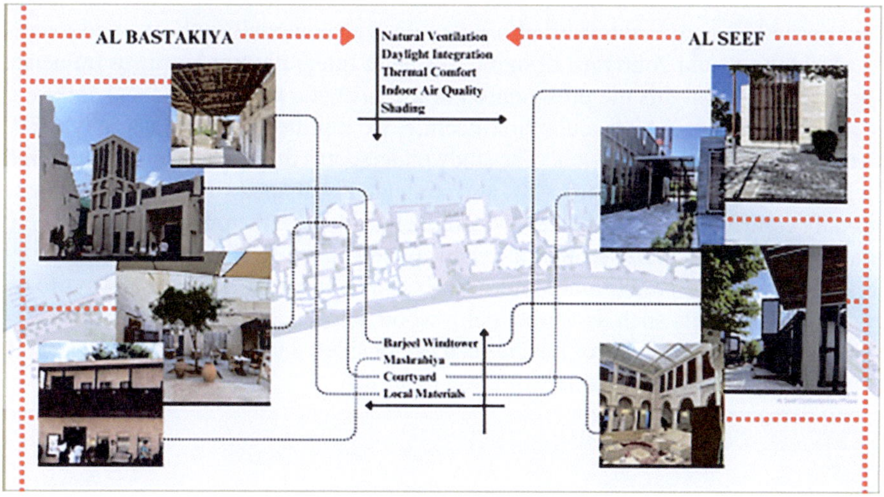

Fig. 21.2 Symbiotic relationship between traditional and contemporary heritage design elements contributing to sustainable interior environment (Al Seef Dubai—Author)

Findings

Research findings from the case study conducted in Dubai, a rapidly growing city, shed light on the transformative impact of Al Seef, a heritage tourist destination of intangible value (Cracknell, 2022). Since 2018, Al Seef has become a treasured space. It is a popular destination with tourists. The public realm design and all the raw materials complement the backdrop of the traditional barjeels (wind towers) and arish (palm-leaf architecture). Shaded areas made from palm leaves (barasti) represented the skill of the historic community who were resourceful with available materials. This locale, characterized by a dual composition featuring old architecture in one section and contemporary structures in another, serves as a cultural bridge to the trading routes of the past. Al Seef revitalizes Old Dubai, offering a culturally enriching experience for both the local community and visitors.

The developed framework resulting from the research has the potential to augment the capacities of designers and scholars. Specifically, it emphasizes the importance of incorporating localities and acknowledging local realities in design strategies, particularly in regions or countries undergoing development or new developments (Fig. 21.2).

Key Research

Ventilation Enhancement Models: The study identifies three models that can be repurposed to improve the ventilation process. These models include wind towers, Al Mashrabiya, and courtyard designs. Aesthetic Integration of Heritage Influence: The research highlights the presentation of the aesthetic effect from local culture as decorative features in vernacular architecture, exemplified by Al Mashrabiya. Efficient Use of Natural Light Tools: The study underscores the efficiency of natural light tools, such as openings, courtyards, and strategically shaped roofs. These elements not only contribute to sustainable design but also alleviate the pressure on electricity consumption, particularly from non-renewable resources.

Adaptive Use of Traditional Materials: Explore innovative and hybrid ways to use traditional materials such as adobe, palm wood, and gypsum to improve indoor air quality. Modern treatments or finishes can be considered to enhance durability while preserving the aesthetic qualities of these materials.

Conclusion

Sustainable architecture is built on the idea of regionalism; however, it does not ask us to return to the old ways of living as pure nostalgia; it inspires us for the responsible, long-term use of technology and design (McLennan, 2004). The developed framework of this study is anticipated to enhance the technical and aesthetic dimensions of designers and scholars by incorporating localities and cultural authenticities into their design strategies.

Significance and Contribution

Theoretical Relevance: This research contributes to the theoretical understanding of the symbiotic relationship between cultural heritage design and sustainable interior design. It expands existing theoretical frameworks by exploring how heritage elements can be practically integrated to enhance the environmental performance of interior spaces.

Practical Relevance: The findings of this study offer practical insights for architects, interior designers, and policymakers involved in the creation of interior spaces considering Al Seef Dubai as an experiential example as the research provides a practical framework for the integration of heritage elements in contemporary design, while seamlessly addressing the need for sustainable design solutions that respect cultural identity.

Limitations

The study focuses specifically on the integration of heritage design elements into sustainable interior environments within the UAE context. Thus, the findings may not be directly applicable to other regions with different cultural contexts or environmental conditions.

However, in conclusion the study recommends that with the suggested framework each region can strategize their design plans according to local and cultural surroundings emphasizing the importance of embracing local culture and heritage design while also employing sustainable design practices in a built environment.

References

Abdelkader, R., & Park, J.-H. (2018). Sustainable building façades: Modern usages of the traditional Mashrabiya. *Open House International, 43*(2), 69–76.

Adam, R. (2012). *The globalisation of modern architecture: The impact of politics, economics and social change on architecture and urban design since 1990.* Cambridge Scholars Publishing.

Al Qasimi, S. K. H. (2019). The impact of implementing heritage elements in contemporary buildings in the United Arab Emirates.

Awad, J., & Boudiaf, B. (2020). Re-using heritage elements in new buildings: Cases from the United Arab Emirates. Southampton.

Chaves, & Alexandra. (2020, 21/12/2023). Arish: the palm leaf's role in the past and future of sustainable architecture in the UAE. *In a Tashkeel Live Lecture, architect Sandra Piesik will discuss the role of vernacular architecture in rebuilding a post-Covid world.*

Cracknell. (2022). Al Seef Heritage Waterfront by Cracknell.

Hanson, N. R. (1965). *Patterns of discovery: An inquiry into the conceptual foundations of science.* CUP Archive.

Hartel, T., Fischer, J., Câmpeanu, C., Milcu, A. I., Hanspach, J., & Fazey, I. (2014). The importance of ecosystem services for rural inhabitants in a changing cultural landscape in Romania. *Ecology and society, 19*(2).
Ibrahim, I. (2019). Eco-traditional courtyard houses in UAE: A case study of the Sharjah museums. *WIT Transactions on the Built Environment, 183*, 15. https://doi.org/10.2495/ARC180021
McLennan, J. F. (2004). *The philosophy of sustainable design: The future of architecture.* Ecotone publishing.
Ragete, F. (2003). *Traditional domestic architecture of the Arab region.* Edition Axel Menges.
Rashdan, W., & Ashour, A. F. (2017). Criteria for sustainable interior design solutions. *WIT Transactions on Ecology and the Environment, 223*, 311–322.
Rashdan, W., & Mhatre, V. (2019). Impact of heritage on contemporary sustainable interior design solutions. *WIT Transactions on Ecology and the Environment, 238*, 47–58.
Rizvi, K. (2017). Dubai, Anyplace: Histories of architecture in the contemporary Middle East. *A Companion to Islamic Art and Architecture*, 1245–1266.
Rudofsky, B. (1987). *Architecture without architects: a short introduction to non-pedigreed architecture.* UNM Press.
Wikipedia contributors. (2024, February 20). Architecture of the United Arab Emirates. In Wikipedia, The Free Encyclopedia. Retrieved14:40, March10, 2024, from https://en.wikipedia.org/w/index.php?title=Architecture_of_the_United_Arab_Emirates&oldid=1209131646
Yahya, N., & Hassanpour, B. (2022). A methodical framework for sustainable architectural design: Housing practice in the Middle East. *Land, 11*(7), 1019.

Chapter 22
Optimizing Cutting Parameters for Achieving Net-Zero Emissions in Dry Machining of Special-Grade Steels

Pradeep George and Pramod George

Abstract In this study, the primary focus is on aligning machining practices with sustainability goals to achieve net-zero emissions. Wet machining, the widely practiced method, involves the use of cutting fluids, which pose significant environmental risks. These risks include the generation of hazardous waste, potential fluid leakage or spillage, soil and water contamination, and adverse health effects through inhalation, skin contact, or ingestion. Moreover, cutting fluids can cause corrosion of machine components and workpieces. To address these issues, an alternative approach known as dry machining is chosen to mitigate environmental impact, reduce costs, and enhance workplace safety. Dry machining requires proper machining parameters to achieve a good surface finish of the workpiece. Considering environmental concerns regarding industrial waste generated during machining, there is a need for greater attention to dry machining. Further exploration into dry machining techniques, especially concerning harder metals like duplex steels, is crucial. Duplex steel (DSS 2205) and super duplex steel (SDSS 2507) were selected for analysis due to their exceptional strength and corrosion resistance properties, particularly advantageous in marine applications. High-speed CNC dry milling of duplex stainless steels with optimum cutting parameters offers substantial sustainability and performance benefits. By eliminating the need for cutting fluids, dry machining reduces environmental impact by minimizing chemical usage and waste generation. Additionally, it can improve surface finish and dimensional accuracy by eliminating coolant-related issues, such as chip evacuation problems or thermal distortion. Furthermore, dry machining enhances productivity by reducing downtime associated with coolant setup and maintenance. Overall, dry machining presents a sustainable and efficient

P. George (✉)
Department of Mechanical Engineering, Energy and Engineering Management, De Montfort University Dubai, Dubai, United Arab Emirates
e-mail: Pradeep.george2@dmu.ac.uk

P. George
Asst. Professor Senior scale, Department of Mechanical Engineering, Manipal Academy of Higher Education Dubai, Dubai, United Arab Emirates
e-mail: pramod.george@manipaldubai.com

© The Author(s), under exclusive license to Springer Nature Switzerland AG 2026
J. Dsilva (ed.), *Redefining Sustainability*, https://doi.org/10.1007/978-3-031-88376-7_22

solution for machining DSS, aligning with net-zero initiatives by reducing resource consumption and environmental footprint.

Keywords Net-zero emissions · Sustainability · Duplex stainless steel · Super duplex stainless steel · High-speed CNC dry milling

Introduction

The use of cutting fluids in wet machining poses significant environmental and health risks, including hazardous waste generation, fluid leakage, and soil and water contamination. Worker exposure to these chemicals can cause respiratory issues, skin irritation, and long-term health effects. Additionally, the costs associated with purchasing, storing, and disposing of cutting fluids, as well as maintenance expenses, add to operational costs. Corrosion of machine components and compliance with regulations further complicate matters, leading to increased downtime and labor costs. Alternative approaches like dry machining are essential to mitigate these issues and improve workplace safety.

Optimizing machining processes for duplex stainless steels (DSS) support net-zero action by reducing energy consumption and minimizing waste generation. By adjusting parameters like feed rate and spindle speed, manufacturers can cut carbon emissions and enhance resource efficiency. Additionally, choosing DSS for their durability contributes to product longevity and supports sustainability goals. Integrating these practices into supply chain initiatives further advances net-zero objectives by minimizing resource extraction and waste throughout the product lifecycle.

Duplex stainless steels like DSS 2205 and SDSS 2507 provide a solution to corrosion concerns in traditional steels. These materials combine ferritic and austenitic characteristics, offering reliability and exceptional corrosion resistance. DSS 2205 blends ferrite and austenite phases, providing a balanced combination of properties, while SDSS 2507, with its higher chromium, nickel, and molybdenum content, offers superior corrosion resistance and strength.

Airao et al. (2018) aimed to investigate how varying cutting speed and feed rate, each having three different values, along with a fixed depth of cut, affect milling operations on 2507 SDSS material, under conditions with and without cutting fluid. Surface roughness resulting from these parameters was analyzed using regression methods. The findings revealed that the feed rate has a more substantial impact on surface roughness compared to the cutting speed.

Wickström et al. (2016) explored duplex stainless steels, commonly employed in coastal environments where susceptibility to corrosion from seawater is high. This research delved into investigating the factors contributing to crack propagation in DSS and SDSS metals. Experimental findings highlighted that crack development was occurring within the nanocrystalline layer formed during the grinding process, alongside the emergence of cracks in the dealloyed layer.

Kulekci (2007) explored dry machining's traits, highlighting easy chip handling and recycling. The method keeps chips uncontaminated as they were separate from cutting fluid, ensuring cleanliness and preventing pollution. Eliminating fluid-related processes notably reduces production costs. High-hardness materials like polycrystalline diamond and ceramics are used, along with coated tools, making dry machining feasible.

In Philip et al. (2018), milling operations on 5A and 4A grade DSS were conducted using WC-coated tools to assess their performance in the machining process. The evaluation of material machinability involved utilizing three different spindle speeds and three feed values, all paired with a constant depth of cut. The results of the experiments demonstrated that an increase in spindle speed led to a decrease in both cutting force and surface roughness values.

Philip et al. (2018) investigated the influence of cutting parameters on surface roughness during turning of duplex steel materials. Five cutting speed variations and three feed values were tested on grade 5A and 4A duplex materials, maintaining a constant depth of cut. Findings indicated that lowering the feed rate reduced surface roughness, while higher cutting speeds resulted in smoother surfaces. Grade 4A displayed a better surface finish due to the presence of α fibers in its austenite region, outperforming grade 5A in surface quality.

Sharma et al. (2022a, 2022b) investigated DSS 2205's machinability for surface roughness, comparing dry versus wet machining using a TiAlN-coated solid carbide drill. Experiments, conducted via CNC milling, employed Taguchi's L8 orthogonal arrays for optimization. Dry machining showed superior surface quality ($R2 = 0.990$) versus wet machining, yielding a finer surface finish. Optimal parameters for desired surface roughness (Ra) were identified as dry environment, high spindle speed (900 RPM), and low feed rate (0.03).

Sharma et al. (2022a, 2022b) studied that dry machining is an environmentally safe alternative to conventional machining. New tools and tool coating (ceramic, diamond, multi-layered) methods have been developed to reduce the cutting heat generation and to extend the tool life. Dry machining is considered a more sustainable process than machining with cutting fluid due to the absence of lubricants and coolants, which results in a reduced use of resources. There are many advantages of dry machining, such as increased flexibility, decreased cycle times, reductions in machine tool costs, and elimination of environmentally hazardous cutting fluids.

Past scholarly works have extensively explored various types of steels and their advancements, shedding light on optimal cutting parameters for diverse machining operations through analytical techniques. However, there are noticeable research gaps, particularly in the realm of dry milling operations involving duplex and super duplex materials, which have received less attention compared to wet machining operations. Given the growing emphasis on net-zero emissions and sustainability, there is an imperative to prioritize dry machining techniques to mitigate environmental concerns associated with industrial waste generated during machining processes. Therefore, there is a need for further exploration into dry machining methods, particularly for harder metals like duplex steels. This necessitates comprehensive

Table 22.1 Chemical composition of DSS 2205 and SDSS 2507

Alloy	Cr	Ni	Mo	C	N	Mn	Si	P	S	Fe
AISI 2205	22.45	5.265	3.081	0.028	0.137	1.632	0.451	0.026	0.006	Balance
AISI 2507	25.38	6.637	3.658	0.024	0.284	0.657	0.370	0.024	0.004	Balance

studies aimed at determining optimal machining parameters to enhance efficiency and sustainability in metalworking practices.

Methodology

The study focuses on optimizing machining parameters for CNC dry milling operations of duplex stainless steel (DSS) to support net-zero action by reducing energy consumption and minimizing waste generation. By adjusting parameters like feed rate and spindle speed, manufacturers can cut carbon emissions and enhance resource efficiency. Additionally, choosing DSS for their durability contributes to product longevity and supports sustainability goals. Integrating these practices into supply chain initiatives further advances net-zero objectives by minimizing resource extraction and waste throughout the product lifecycle.

Work Material and Tool Selection

This study focuses on aligning machining practices with sustainability goals to achieve net-zero emissions. DSS 2205 and SDSS 2507 materials were chosen for experimentation. Table 22.1 outlines their chemical properties. Workpiece measuring 100 mm in length, 80 mm in width, and 50 mm in thickness were chosen for milling operations. The BMV 40 T20 CNC machining center, equipped with a spindle speed range of 60 to 6000 rpm and a 10 kW spindle motor rating, was used for experimentation. Ø16 mm end milling cutters with specifications EAP300R-16D-160L-2T-W and TiCN-coated tungsten carbide inserts APMT 1135PDR-YBG 205 were selected for the investigation. Surface roughness was evaluated using a Mitutoyo (Model SJ210) roughness tester.

Cutting Parameters Selection

In this experiment, three parameters were considered for the dry milling of DSS 2205 and SDSS 2507 metals using a CNC milling machine. Cutting speed ranging from

Fig. 22.1 Experimental setup containing CNC milling machine and measuring devices

1400 to 5600 rpm was used. The feed rate was selected between 50 and 150 mm/min, and a constant depth of cut of 0.7 mm was maintained throughout the operation. The experiments were conducted based on the design matrix. The effect of cutting parameters on surface roughness was estimated. The arrangement of the experiment is shown in Fig. 22.1.

Results and Discussions

The results and discussion of CNC dry machining of duplex steel offer valuable insights into the feasibility and effectiveness of this machining approach, particularly concerning its environmental impact and resource efficiency (Table 22.2).

Impact of Machining Parameters on Surface Roughness

Figures 22.2 and 22.3 depict how spindle speed impacts surface roughness in high-speed dry milling of DSS 2205 and SDSS 2507 at varied feed rates (50–150 mm/min) and cutting speeds (1400–5600 rpm). The data highlights that the lowest roughness occurs at 4200 rpm.

Figures 22.4 and 22.5 showcase how feed rate impacts surface roughness during high-speed dry milling of DSS 2205 material across four cutting speed levels. The data indicates that the lowest roughness occurs at a feed rate of 50 mm/min.

Table 22.2 Experimental results for surface roughness of DSS 2205 and SDSS 2507

Expt. No.	Spindle speed N (rpm)	Feed rate, f (mm/min)	Surface roughness, Ra (μm)	
			DSS 2205	SDSS 2507
1	1400	50	0.68	0.87
2	2800	50	0.57	0.83
3	4200	50	0.52	0.81
4	5600	50	0.76	0.97
5	1400	100	0.81	0.95
6	2800	100	0.78	0.95
7	4200	100	0.66	0.88
8	5600	100	1.09	1.29
9	1400	150	1.01	1.25
10	2800	150	0.91	1.14
11	4200	150	0.86	1.1
12	5600	150	1.74	2.05

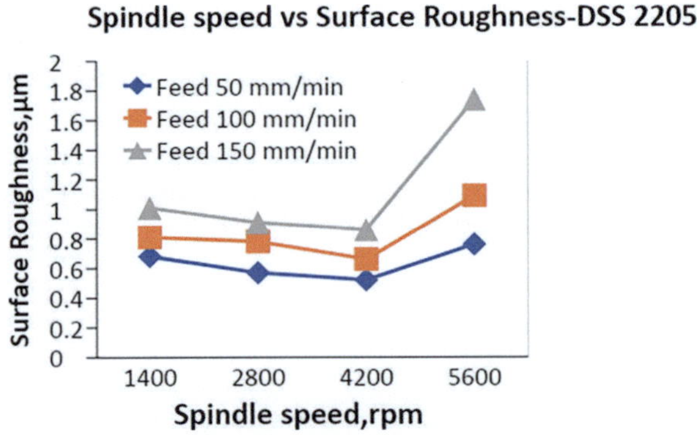

Fig. 22.2 Spindle speed versus surface roughness of grade DSS 2205

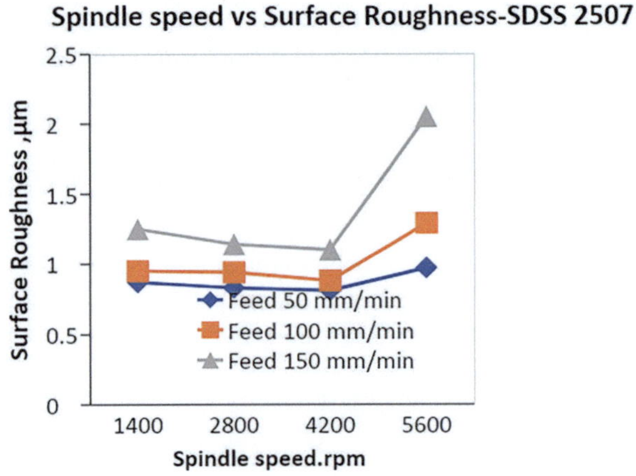

Fig. 22.3 Spindle speed versus surface roughness of grade SDSS 2507

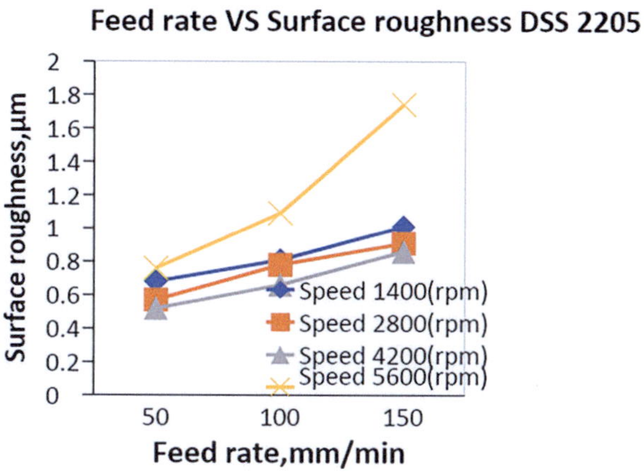

Fig. 22.4 Feed rate versus surface roughness of grade DSS 2205

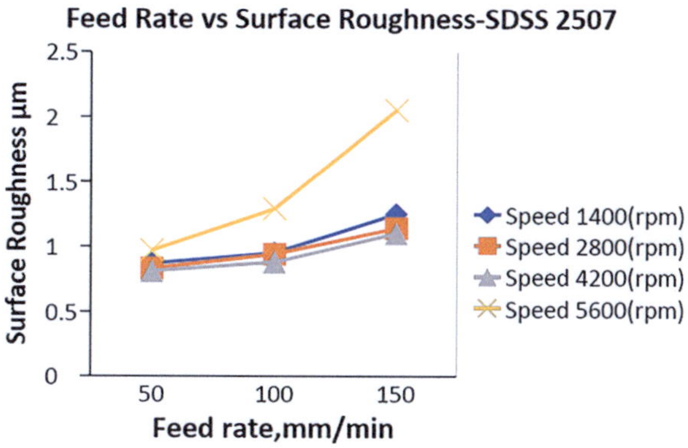

Fig. 22.5 Feed rate versus surface roughness of grade SDSS 2507

Optimum Machining Parameters for CNC Dry Machining of Hardened Steel

The optimum cutting speeds and feeds for dry machining of hardened materials were determined through experimental analysis and was shown in Table 22.3. This minimized energy consumption and reduced machining time, improved surface quality contributing to lower carbon emissions and improved resource efficiency.

Table 22.3 Optimum cutting conditions for surface roughness

Spindle speed, N (rpm)	Feed rate, f (mm/min)	Surface roughness, R_a (µm)	
		DSS 2205	SDSS2507
4200	50	0.52	0.81

Conclusions

- The CNC dry machining of duplex steel demonstrated promising results in terms of surface finish and dimensional accuracy. The absence of coolant minimized the risk of thermal distortion and improved the precision of the machined components.
- The elimination of coolant fluids reduced the need for energy-intensive coolant systems and minimized water usage, contributing to overall energy savings and sustainability.
- From the experiment, it is found that surface roughness for DSS 2205 and SDSS 2507 grade were minimum at 4200 rpm (cutting speed) and 50 mm/min (feed rate).
- Overall, the results and discussion of CNC dry machining of duplex steel highlight its potential as a sustainable and cost-effective alternative to traditional wet machining methods. By leveraging advanced machining techniques and optimizing process parameters, manufacturers can enhance productivity, reduce environmental footprint, and achieve long-term sustainability goals in metalworking operations.

References

Airao, J., Chaudhary, B., Bajpai, V., & Khanna, N. (2018). An Experimental study of surface roughness variation in end milling of super duplex 2507 stainless steel. *Materials Today Proceedings*, *5*(2), 3682–3689.

Kulekci, M. K. (2007). Process and apparatus developments in dry-machining applications. *Journal of Mechanical Engineering*, *53*(9), 541–547.

Philip Selvaraj, D., & Chandramohan, P. (2010). Influence of cutting speed, feed rate and bulk texture on the surface finish of nitrogen alloyed duplex stainless steels during dry turning. *Scientific research Engineering*, *2*, 453–460.

Philip Selvaraj, D, Chandramohan, P, & Chandrasekar. (2018). Experimental investigations of nitrogen alloyed duplex stainless steel in dry milling process.journal of engineering science and technology *13*(2), 321–331.

Sharma, R., Jha, B. K., Pahuja, V., & Sharma, S. (2022a). Role of environmental friendly machining on machinability *50*(5), 640–648.

Sharma, R., Pahuja, V., Jha, B.K., & Sharma, S. (2022b). Evaluation of machinability on duplex stainless steel 2205 under dry and wet condition *65*(8), 3357–3361.

Wickström, L., Mingard, K., Hinds, G., & Turnbu, A. (2016). Micro crack clustering in stress corrosion cracking of 22Cr and 25Cr duplex stainless steels. *Corrosion Science, 109*, 86–93.

Chapter 23
Synergizing Sustainability by Hybridizing Aquaponics for Energy Saving and Food Production in Warm Climate Residential Environments: BaityKool (BK), a Bio-Inspired Dwelling Prototype in Dubai-UAE

V. P. Mohanan, R. Bouzouidja, F. Garcia, J. Locke, J. Dsilva, A. Rochaud, B. Canigiani, P. Lagiere, A. Samuel, and D. Bruneau

Abstract Due to rising temperature and the fluctuations in the weathers condition, a severe crisis is faced by the agriculture industry around the world. Agriculture being the largest consumer of fresh water globally upto 70%, long-standing farming methods which was sturdy and dependable, now encounter numerous challenges jeopardizing global food security. Issues such as imminent water shortages and the gradual depletion of fertile soils present significant threats to traditional agriculture. Consequently, the conventional farming sector finds itself at a crucial juncture, compelling the adoption of innovative, resource-conscious techniques that can adapt to changing conditions. The aquaponics system developed in the Baitykool lab represents an initial move toward utilizing existing surfaces and sunlight, contrasting with interior-controlled environments that consume space and energy for air conditioning and lighting. Aquaponics, widely recognized for large-scale food production, holds significant potential as a farming solution. This research's importance lies in

V. P. Mohanan (✉) · R. Bouzouidja · F. Garcia · P. Lagiere · A. Samuel · D. Bruneau
University of Bordeaux, CNRS, Bordeaux INP, Talence, France
e-mail: vmohan@amityuniversity.ae

Arts Et Metiers Institute of Technology, CNRS, Bordeaux INP, Hesam Universite, Talence, France

V. P. Mohanan · A. Samuel
Amity University Dubai, Dubai, United Arab Emirates

V. P. Mohanan · F. Garcia · A. Samuel · D. Bruneau
GRECCAU, ENSAP Bordeaux, Talence, France

J. Locke · J. Dsilva
SEE Institute, Dubai, United Arab Emirates

A. Rochaud · B. Canigiani
Arkiturria, Montbonnot-Saint-Martin, France

© The Author(s), under exclusive license to Springer Nature Switzerland AG 2026
J. Dsilva (ed.), *Redefining Sustainability*, https://doi.org/10.1007/978-3-031-88376-7_23

advancing the integration of aquaponics directly into the structural walls of residential buildings. A prototype home called BaityKool won the "sustainability" prize at the inaugural Solar Decathlon Middle East competition in November 2018 in Dubai. The research entails a detailed evaluation of BaityKool's aquaponic system efficiency, energy saving capacity, and food yield within Dubai's warm climate. This approach not only signifies a sustainable solution for energy and food production but also serves as an educational platform. The outcomes of this research hold promise for transforming residential design in hot dry regions. By addressing food security, promoting energy sustainability, and advocating for environmental conservation, BaityKool represents a significant stride toward fostering a more sustainable living.

Keywords Sustainability · Aquaponics · Food security · Climate change · Food production · Energy saving · Climate change · Agriculture

Introduction

The demographic composition of the United Arab Emirates (UAE) has witnessed a significant escalation in recent years, expanding from 1.5 million individuals in the 1970s to an excess of 9.5 million by the year 2023 (Alsharhan & Rizk, 2020), a phenomenon attributable to extensive advancements in both infrastructure and economic endeavors. This surge has intensified the strain on the already limited domestic food production capabilities within the UAE, resulting in an increase in annual food imports to over 85% of total consumption (Al Qaydi, 2016), thereby necessitating the exploration of alternative food production strategies that mitigate carbon emissions. The challenge of food security and production emerges as a primary concern in urban contexts where land resources are severely constrained, yet food demand remains exceptionally high (Den Boer et al., 2021). The objective of this study is to assess the viability of aquaponics as a sustainable approach to enhancing family food security within the UAE. This investigation employs a quantitative methodology to evaluate the feasibility of substituting a concrete wall with an aquaponics system capable of yielding a diverse array of vegetables and fish, thereby providing a balanced source of protein and nutrition for the residents inhabiting the space.

The aquaponics system developed by the BaityKool lab represents a groundbreaking initiative, leveraging existing surfaces and sunlight instead of an interior-controlled environment that consumes space and energy for air conditioning and lighting. Widely recognized for large-scale food production, aquaponics holds promise for structural integration into residential buildings. The study aims to highlight the integration of aquaponic systems within building designs, particularly focusing on arid regions like Dubai, UAE.

BaityKool, an award-winning prototype home honored for sustainability at the Solar Decathlon Middle East competition in Dubai (November 2018) (Alshikh et al., 2023), embodies a global bio-inspired design aimed at maximizing water efficiency and enhancing water recycling. The research focuses on incorporating the aquaponics system to provide environmentally friendly food for all BaityKool occupants. This system, compared to conventional vegetable production, demonstrates an impressive energy-saving potential of up to 50%, while contributing to the sub-production of food consumed. The BaityKool prototype is a testament to sustainability, employing advanced technological and bio-inspired solutions to harness natural resources efficiently. The prototype harnesses solar energy for electricity while utilizing fish waste to fertilize plants, creating a symbiotic relationship between energy production and food cultivation (Hosseinian et al., 2023).

The research contributes valuable insights to advance the integration of renewable energy and agricultural practices in residential architecture, fostering a more self-sufficient and environmentally conscious approach to habitation in warm climate. This research explores the integration of aquaponics as a multifunctional system within residential architecture in warm climates, focusing on energy generation and food production.

A comprehensive during the extreme season assessed the aquaponics system's efficiency in terms of water conservation and food yield (both fish and vegetables). Scientific experiments delved into quantifying food production and energy consumption parameters. Conducted in extreme weather conditions of high temperature and humidity, this study presents both the technological design of the aquaponic system and its architectural integration into residential spaces. The findings demonstrate the viability of combining aquaponics for food production with energy generation in warm climate residential environments, showcasing BK as a pioneering model for sustainable living (Abusin & Mandikiana, 2020).

Methodology

BaityKool is an architectural structure designed to accommodate a household of four individuals, specifically tailored for regions characterized by elevated temperatures. The U-shaped layout creates an internal courtyard, serving as a transitional zone that mitigates the extreme external climatic conditions while enhancing the comfort levels within the residence. Its operational mechanisms are predicated upon low-tech methodologies and fundamental principles, such as gravitational forces, which facilitate the movement of water throughout the system (Samuel et al., 2020). An aquaponics system has been seamlessly incorporated into one of the walls. The 1150L acrylic aquarium, embedded within the wall, functions as a transparent interface between the patio and an internal corridor. As the aquarium's water level ascends, overflow occurs at two distinct points located on the periphery of the fish tanks, directing the water toward gravitational biofilters that house the majority of the bacterial colonies. Following this filtration process, the water is channeled through 18 ZipGrow towers, which utilize a vertical nutrient film technique, as illustrated in Fig. 23.1. Each ZipGrow tower consists of a PVC tube with a square cross-section measuring 10 cm, featuring a vertical aperture on one side. Within the tower resides

Fig. 23.1 BaityKool aquaponic system. Constructed with timber cladding to fulfill experimental requiremnts. V. Mohanan©

a specialized fibrous medium known as MatrixMedia, which serves as an anchoring substrate for the plant roots. A textile strip embedded within the MatrixMedia functions to retain both water and nutrients, preventing excessive drainage through the tower. All towers are positioned behind wooden cladding, with the planks arranged to allow for a 3 cm gap, thereby facilitating the growth of plants through the cladding (Mansouri et al., 2009).

A drainage system collects the precipitation that accumulates beneath the edifices and channels it toward the subterranean sump tank, which is discreetly located beneath the terrace. From the sump tank, a submerged pump facilitates the ascension of the water. Prior to its return to the aquarium, the water passes through a ultraviolet (UV) lamp, effectively eliminating pathogens and superfluous bacteria.

As illustrated in Fig. 23.2, the entire volume of water traverses the system at a flow rate of 7500 L/h, facilitated by a singular pump in conjunction with the ultraviolet (UV) lamp, which operates at an energy consumption rate of 4.16 kWh/day. The system accommodates a maximum of seventy-two plants; however, it possesses the capacity to integrate a diverse assortment of fruit-bearing plants, given that the aquarium is oversized in relation to the available planting slots: in comparison with the quantity of plants cultivated within the NeighborHub system, the BaityKool aquarium boasts a total volume that is more than double that of the fish tanks utilized in the NeighborHub setup. Consequently, this allows for a greater number of fish, which in turn results in elevated levels of ammonia, nitrites, and nitrates, thereby augmenting the demand for plants to assimilate these nutrients.

In the course of the performance assessment, the aquarium was populated with 16 fish (weighing 150 g each), specifically Malawi Cichlids. This species was selected

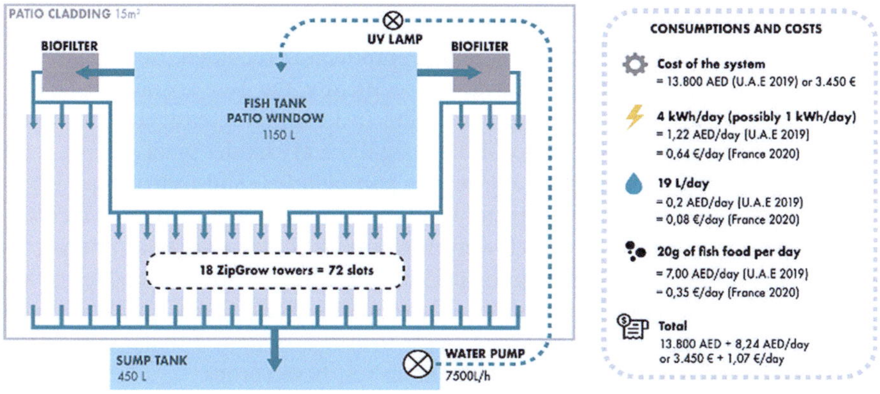

Fig. 23.2 System and consumption data

due to its substantial rate of excreta production. The fish were provided with 20 g of fish feed on a daily basis. The ZipGrow towers were populated with a total of 25 plants: 5 mints, 5 cilantro, 5 tomatoes, 5 chili peppers, and 5 basil plants. The plants were sourced from the nursery prior to their introduction into the aquaponic system and exhibited rapid adaptation to the vertical growth conditions within a few hours, with the stems fortifying themselves and orienting upwards once again (Kang et al., 2015).

Findings

In order to assess the efficacy of the system, three metrics were employed: water utilization (L/day), electrical energy consumption (kWh/day), and agricultural output (Ochoa & Capeluto, 2008).

The system exhibited an electrical consumption of 5 KWh, with water usage for the entire system, including global leaks, quantified at 10–15 L per day (Fig. 23.4). The relationship between fish growth and water temperature is illustrated in Fig. 23.3, which produced a value of 1.2 g/day, contrasting with a rate of 3 g/day observed in traditional aquaculture practices (Fig. 23.4). The tomato plants cultivated within the aquaponic system displayed a growth rate of + 190%, closely resembling the + 215% increase observed in soil-based agricultural practices. Nevertheless, neither the tomato plants from the aquaponics nor those from the soil-based cultivation exhibited flowering or fruiting, as they had not attained maturity by the termination of the experiment. Both chilies and mint demonstrated remarkable acclimatization to the climatic conditions, particularly flourishing in the aquaponic environment. With a notable growth of + 550% within a three-week period and the appearance of green fruits—ranging from 4 to 8 per plant and measuring up to 2 cm in length—at 4.5 weeks, these crops derived substantial benefits from the aquaponic system.

In contrast, their growth rate within conventional cultivation was merely half that observed in the aquaponic setup, with no fruit production recorded. Both basil cultivations faced infestations from parasites—specifically, the Tuta absoluta, a species of fly that lays eggs on basil leaves, resulting in larvae progressively consuming the foliage. Although this infestation did not significantly hinder plant growth, it did diminish the availability of consumable leaves, particularly in soil-based crops. Under standard conditions, basil plants typically yield greater quantities than those recorded in this experiment. The observed slower growth is attributed to regular pruning for self-consumption or to the removal of parasites, leading to a more productive and denser basil crop. When considering equivalent plant development, aquaponic basil enabled a harvest that was eight times larger than that of conventional cultivation: yielding 2.5 heads per plant compared to 0.3 in soil-based crops.

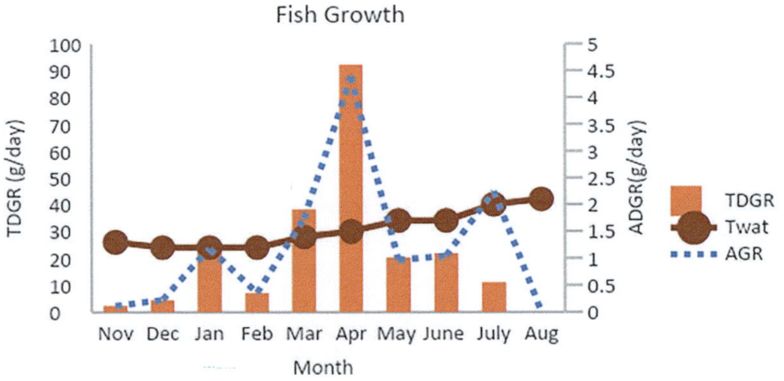

Fig. 23.3 Growth rate of fish over the season

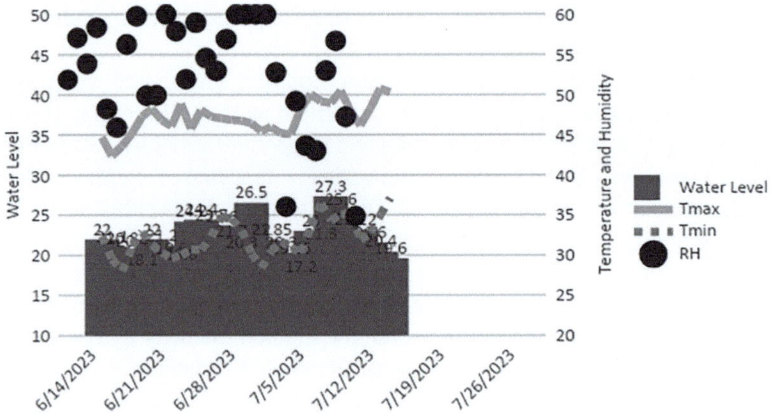

Fig. 23.4 Water consumption of the aquaponic system

Conclusion

Aquaponics presents an appealing solution for tackling food security challenges in urban areas constrained by limited land availability. This system allows for the cultivation of agriculture within small spaces, offering a straightforward, convenient, and effective approach. In the initial four-week duration subsequent to the implementation of the aquaponics system, this innovative methodology demonstrates a markedly greater efficacy than the traditional culture method. The global development, as evidenced by the mean plant growth metrics, displays a remarkable 15-fold enhancement within the aquaponic system (with an impressive growth rate of 150%) in comparison with the conventional culture system (which registered a growth rate of merely 10%). It is crucial to acknowledge that while the experimental findings indicate a superior efficiency in terms of growth rates, these results do not inherently signify an elevated level of food production, given that the plants had not attained maturity throughout the span of the four-week experimental timeframe. The numerous benefits of the system, such as minimal maintenance costs, high productivity, efficient water usage, elimination of harmful pesticides, and reduced land usage, position it as an attractive option for addressing the food requirements of urban households with restricted land resources. Moreover, the adoption of aquaponics not only contributes to meeting food needs but also has positive environmental impacts by enhancing visual aesthetics and maintaining natural ecosystem equilibrium.

References

Abusin, S. A. A., & Mandikiana, B. W. (2020). Towards sustainable food production systems in Qatar: Assessment of the viability of aquaponics. *Global Food Security, 25*, 100349. https://doi.org/10.1016/j.gfs.2020.100349

Alsharhan, A. S., & Rizk, Z. E. (2020). *Water resources and integrated management of the United Arab Emirates* (Vol. 3). Springer International Publishing. https://doi.org/10.1007/978-3-030-31684-6

Alshikh, Z., Trepci, E., & Rodriguez-Ubinas, E. (2023). Sustainable off-site construction in desert environments: Zero-energy houses as case studies. *Sustainability, 15*(15), Article 15. https://doi.org/10.3390/su151511909

Al Qaydi, S. (2016). The status and prospects for agriculture in the United Arab Emirates (UAE) and their potential to contribute to food security. *Journal of Basic & Applied Sciences, 12*, 155–163. https://doi.org/10.6000/1927-5129.2016.12.23

Den Boer, A. C., Broerse, J. E., & Regeer, B. J. (2021). The need for capacity building to accelerate food system transformation. *Current Opinion in Food Science, 42*, 119–126. https://doi.org/10.1016/j.cofs.2021.05.009

Hosseinian, S. M., Sabouri, A. G. A., & Carmichael, D. G. (2023). Sustainable production of buildings based on Iranian vernacular patterns: A water footprint analysis. *Building and Environment, 242*, 110605. https://doi.org/10.1016/j.buildenv.2023.110605

Kang, J. E., Ahn, K. U., Park, C. S., & Schuetze, T. (2015). Assessment of passive versus active strategies for a school building design. *Sustainability, 7*(11), Article 11. https://doi.org/10.3390/su71115136

Mansouri, A., Dowdle, S., & Dowdle, D. (2009). *Climatic design strategies in hot regions: An application of Tripoli-Libya.*

Ochoa, C. E., & Capeluto, I. G. (2008). Strategic decision-making for intelligent buildings: Comparative impact of passive design strategies and active features in a hot climate. *Building and Environment, 43*(11), 1829–1839. https://doi.org/10.1016/j.buildenv.2007.10.018

Samuel, A., Mohanan, V., Sempey, A., Yusta Garcia, F., Lagiere, P., Bruneau, D., & Mahanta, N. (2020). *A self-sustainable home: 'BAITYKOOL' developed for the extreme warm climate an experimentation of active & passive strategies* (p. 7). https://doi.org/10.1109/ASET48392.2020.9118326

Chapter 24
The Joyce Centre—Design and Performance Metrics of a Certified Zero-Carbon Institutional Building

Anthony F. Cupido

Abstract There is limited measured energy performance knowledge generated from certified zero-carbon institutional buildings in Canada. The objective of this study is to compare as-built energy and performance models with operational energy consumption and energy generation data to validate if Canada's first dual certified (design and performance) zero-carbon building (ZCB) is operating as intended, and to determine the cause of any variances between the modeled energy values and the measured energy consumption values of independent variables. A formal review of design methodologies, technical specifications and reports, energy targets and metrics was undertaken and summarized. Building operations data and generated renewable energy data collected over the first four years of operations consisting of energy use intensity (EUI), thermal energy demand intensity (TEDI), process loads, solar photovoltaic (PV) generation, and solar thermal domestic hot water (DHW) production has been compared to as-built modeled values and design targets of the study building. The findings of this study demonstrate the building has outstanding performance metrics and is operating as a net zero-energy and zero-carbon building. The average EUI is 65% of the targeted design EUI of 72 eKwh/m^2-yr. and the building energy consumption is approximately 66% of the PV energy generation. However, independent variables, consisting of occupancy-based loads, have contributed to reduced energy consumption and an over-designed PV array capacity. This has resulted in a significant difference between several modeled energy values versus consumption energy values. Inaccurately modeled variables in a ZCB may lead to unintended consequences such as higher energy production than necessary and associated higher costs, which occurred with the study building.

Keywords Zero-carbon building · EUI · Renewable energy · TEDI · Independent variables

A. F. Cupido (✉)
Research Chair, Sustainable Building Technologies—Mohawk College, Hamilton, Canada
e-mail: tony.cupido@mohawkcollege.ca

© The Author(s), under exclusive license to Springer Nature Switzerland AG 2026

J. Dsilva (ed.), *Redefining Sustainability*, https://doi.org/10.1007/978-3-031-88376-7_24

Introduction

Leading industry authorities have acknowledged that high performance buildings need to consider design, construction and operational aspects to ensure the building's sustainable goals are met (ASHRAE, 2010). The Canada Green Building Council (CaGBC) has defined a zero-carbon building as highly energy efficient and produces on-site carbon-free renewable energy in an amount sufficient to offset the annual carbon emissions associated with the building materials and operations (CaGBC, 2020). The 8981 m^2 study building, formally known as the Joyce Centre for Partnership and Innovation, integrated a high-efficiency design, construction materials and technologies, and renewable energy technologies to significantly reduce its annual energy consumption and targeted zero GHG emissions (Bhavsar et al., 2020).

The problem, however, is due to the limited institutional participation in the CaGBC's ZCB V2 Program, there is a lack of operational energy performance knowledge of performance certified ZCB's in Canada. Some modeled energy values in the study building also varied significantly from the measured energy consumption values, which has led to questions regarding the modeled values of independent variables (lighting, plug loads, process loads, and DHW), occupancy rates, and PV array performance. The objective of this study is to compare as-built energy and performance models with operational energy consumption and energy generation data to validate if Canada's first dual-certified (design and performance) institutional ZCB under the CaGBC's ZCB Program is operating as intended, and to determine the cause of variances between the modeled energy values and the measured energy consumption values of independent variables.

A study by Bui et al. (2023) recognized that challenges and barriers exist to zero carbon buildings, namely the shortage of knowledge, capacity, and capability related to whole-of-life building carbon reduction. Maduta et al. (2022) acknowledged the importance of numerical benchmarks considering both energy and carbon metrics in zero emission buildings along with GHG calculation over the building life-time. This study also recognized that focus has been on estimated and actual operational energy use, but efforts should be made on controlling the energy use of heating, cooling, ventilation, hot water, lighting, and auxiliary energy. In a study of carbon and energy metrics for net-zero buildings by Parkin et al. (2020), a framework was developed that describes a building as a system of interacting components that include building energy demand profiles, PV energy generation profiles and properties of the building itself such as the thermal envelope. The framework included often disregarded components such as occupant energy demand. Carpino et al. (2020) recognized that user interaction in high-efficiency buildings may result in a total energy use that is considerably different than the estimated performances. The accuracy in the description of occupancy ultimately affects the reliability of energy performance predictions.

Methodology

The following methodology and approach were used to inform the conclusions of this study. Data from end use building components such as lighting, process and plug loads, space heating and cooling, and fans were identified and modeled by the design professionals, and available for this study. A metering and submetering complement, consisting of 48 devices, measured each end use component with an emphasis on lighting and process loads. Building occupancy projections were developed during the functional planning of the facility and were originally estimated to be 1,100 occupants. This study did not include measurements related to embodied carbon or refrigerants.

In addition to the considerable volume of planning, design, construction, and commissioning documentation available to confirm energy targets and predicted occupancy, the author of the study has collected primary data from the study building that includes annual energy consumption data since summer 2018 from the digital metering system (DMS), the building automation system (BAS), and campus utility meters. Fifteen-minute accumulated consumption data (kWh) was analyzed and monthly totals for each submeter were obtained from the accumulated energy data by subtracting the start of month from the end of month. Energy data generated by the building end uses was collected locally at the Facility Service's central utility plant and used for this study. All energy consumption data gathered was analyzed and compared to modeled information available.

Findings and Discussion

Since the opening of the facility in September 2018, the building energy performance has been exceptional and continues to exceed expectations. The following energy performance metrics have validated the exceptional performance over the last 4 years:

- The average EUI is 65% of the targeted design EUI of 72 eKwh/m^2-yr.
- The design TEDI of 22 ekWh/m^2-yr. has been met.
- The building is generating, on average, approximately 35% more electrical energy than it consumes annually.
- The seasonal peak demand occurs in the winter as anticipated for the geographic location.
- The solar thermal system performs adequately but has a high ROI.
- IT server loads and laboratory plug loads are two of the largest process loads as expected, however, IT electrical loads are significantly lower than the designed capacity by almost 190,000 kWh.

The energy consumption of independent variables such as lighting, plug loads, and process loads (inclusive of IT loads) are largely occupancy and user driven. Independent variables have a measurable impact on energy consumption and modeled values varied significantly from measured values by as much as 70% for IT electrical loads for example. The IT estimated loads related to volume of equipment and cooling needs determined at the time of functional planning appear to have been overestimated by IT staff. A legitimate argument can be made that the on-site roof-mounted photovoltaic system consisting of 1980 photovoltaic panels with a capacity of 550 kWp AC and an annual generation of 721,000 kWh has a significant surplus capacity. A conservative design reduction of 390 PV panels is warranted with considerable savings on supply and installation costs.

Conclusions

Achieving net zero energy and zero carbon requires a holistic building design strategy starting with mitigating energy loads as much as possible. Establishing a target EUI at the onset of the design was a critical factor for achieving performance requirements and guided designers to meet those targets and operators to adhere to them. The measurable impact of independent variables may require more rigor at the design stage to determine more accurate (or more realistic) modeled values for energy use and building occupancy. Inaccurately modeled variables in a ZCB may lead to unintended consequences such as higher energy production than necessary and associated higher installation costs. This study has demonstrated a successful building performance and compliance with the CaGBC's ZCB Performance Standard V2. This study and the data generated will be invaluable to the Canadian building industry, design professionals, researchers, students, and the green building industry globally. The Joyce Centre institutional building offers a distinct opportunity to advance concepts and knowledge around zero-energy and zero-carbon technologies at a regional, provincial, national, and international level.

In spite of these conclusions, there are potential limitations in this study. The author of this study is a member of the Board of the CaGBC and some bias may be perceived. The study sample size was limited to one facility and additional facilities are needed to improve any industry dataset. The COVID pandemic created variances in occupancy, inclusive of a full building closure to asynchronous approaches to learning, that have impacted on the daily occupancies of staff and students.

Recommendations for future research include integrating new data from additional performance certified facilities, undertaking a formal occupancy study to compare design targets to measured daily occupancy throughout the entire year, and having college facility staff better trained to measure and identify independent variable irregularities.

References

ASHRAE. (2010). ASHRAE greenguide: The design, construction, and operations of sustainable buildings. In J. Swift, T. Lawrence, A. Pavlik, M. Owen, C. Michaels, J. Walker, & M. Phillips, (Eds.). ASHRAE Special Publications.

Bhavsar, R., Cupido, A., & Arriaga, M. (2020). Mohawk college's net zero energy and zero carbon building—a living lab for high efficiency and renewable energy technologies in buildings. *Journal of Green Building, 15*(1), 185–214.

Bui, T. T. P., MacGregor, C., Wilkinson, S., & Domingo, N. (2023). Towards zero carbon buildings: Issues and challenges in the New Zealand construction sector. *International Journal of Construction Management, 23*(15), 2709–2716.

Canada Green Building Council. (2020). *Zero carbon building performance standard version 2.* Retrieved January 15, 2024, from the Canada Green Building Council: https://www.cagbc.org/ourwork/certification/zero-carbon-building-standard/zero-carbon-building-standard-performance-resources/

Carpino, C., Loukou, E., Heiselberg, P., & Arcuri, N. (2020). Energy performance gap of a nearly Zero Energy Building (nZEB) in Denmark: The influence of occupancy modelling. *Building Research and Information, 48*(8), 899–921.

Maduta, C., Melica, G., D'Agostino, D., & Bertoldi, P. (2022). Towards a decarbonised building stock by 2050: The meaning and the role of zero emission buildings (ZEBs) in Europe. *Energy Strategy Reviews, 44*, 101009.

Parkin, A., Herrera, M., & Coley, D. A. (2020). Net-zero buildings: When carbon and energy metrics diverge. *Buildings and Cities, 1*(1), 86–99.

Chapter 25
Experimental Investigation on the Flexural Behavior of Steel-Timber Composite Floor Panels

Seifaldeen Dandashi, Gul Ahmed Jokhio, and Fadi Elayache

Abstract The carbon footprint of buildings primarily originates from the production of abiotic building materials like concrete, metals, and plastics, accounting for approximately 14% of global greenhouse gas emissions. Thus, selecting low-carbon materials such as timber becomes increasingly crucial. This paper aims to evaluate the impact of steel-timber composite floor panels on reducing the overall carbon footprint of buildings. Two different structural schemes were compared. The findings indicate that using steel-timber floor panels can decrease the overall carbon footprint by one third compared to traditional reinforced concrete structures. Additionally, while the composite behavior of conventional steelwork and concrete slabs is well-documented and integrated into international building codes, the exploration and standardization of steel-timber composites are still in progress. This study concentrated on the effect of shear connector spacing on the bending behavior of the specimens, with results evaluated in comparison with similar studies. The paper presents the study's outcomes, revealing that shear connector spacing significantly impacts bending behavior, achieving up to a 48% increase in flexural stiffness normalized by the non-composite stiffness. Furthermore, the moment capacity of composite sections was discovered to be five times greater than that of bare steel sections.

Keywords Sustainable built environment · Embodied carbon · Composite steel-timber floor · Bending stiffness · Flexural behavior of floor panels · Flexural stiffness of composite floor panels

S. Dandashi (✉) · G. A. Jokhio
The British University in Dubai, Dubai, United Arab Emirates
e-mail: dandashi.seif@gmail.com

G. A. Jokhio
e-mail: gul.ahmed@buid.ac.ae

F. Elayache
Inventive Modular, Dubai, United Arab Emirates
e-mail: fadi@inventivemodular.com

© The Author(s), under exclusive license to Springer Nature Switzerland AG 2026
J. Dsilva (ed.), *Redefining Sustainability*, https://doi.org/10.1007/978-3-031-88376-7_25

Introduction

Although the composite action for conventional steelwork and concrete slabs is well understood and documented into international building codes, this advantage has yet to be well explored and standardized for steel-timber composites. Several studies have been conducted in the past few years to describe the benefits of timber panels on steel joists quantitatively. For example, Kyvelou et al. (2015) assessed the cold-formed steel joists supporting particleboards under bending. The assembled system reached partial shear connection up to 68% of full interaction and resulted in as much as a 100% increase in bending capacity and a 42% increase in stiffness. Full-scale bending and push-out tests on composite specimens made of cold-formed steel joists and wood-based floorboards adjoined using screws and adhesive were conducted by Raffoul et al. (2019), and the improvements found in the composite action were in good agreement with the study conducted by Kyvelou et al. (2015). In the experiment conducted by Loss and Davison (2017), two prototypes were fabricated and assembled using different methods of shear connection. In the first specimen, the CLT panel was assembled with the customized steel profile using a combination of inclined fully threaded screws and perpendicular self-drilling screws, while the second specimen relied on the equally spaced steel plate inserted with adhesive into grooves in the CLT. The flexural performance of the specimens was ductile, and the achieved bending stiffness (EI) in the first specimen was two times higher than the stiffness of the steel when, in fact, the second specimen achieved a bending stiffness nearly five times higher than the bare steel joists. To the best of our knowledge, the flexural performance of composite floor panels made of cold-formed steel (CFS) and laminated veneer lumber (LVL) has not been investigated yet. Therefore, experimental four-point loading testing on three full-scale specimens was conducted. The results were assessed in the light of similar studies.

Experimental Program

Material Properties: The material properties used for the steel joists, the shear connectors, and the LVL panels are presented as follows.

Steel Joists: Six steel joists with a 4.2 m span were fabricated using roll formers at Baghalf Al Zafer Metal Industries from S350 zinc-coated galvanized steel coils. The joist profile consisted of a single symmetric lipped C-shaped section of a nominal depth of 153 mm, flange width of 37 mm, lip depth of 15 mm, and thickness of 1.5 mm, as shown in Fig. 25.1. The other properties of the steel joists are listed in Table 25.1.

Shear Connectors: Self-tapping screws JA2-6.5 manufactured by EJOT were used as shear connectors. This type of screw is used to fix thick metal sheets on timber substructures. The thread and root diameter were 6.5 and 4.7 mm, respectively. The length of the screw was 50 mm, and the height of the hexagon head was 5 mm. The

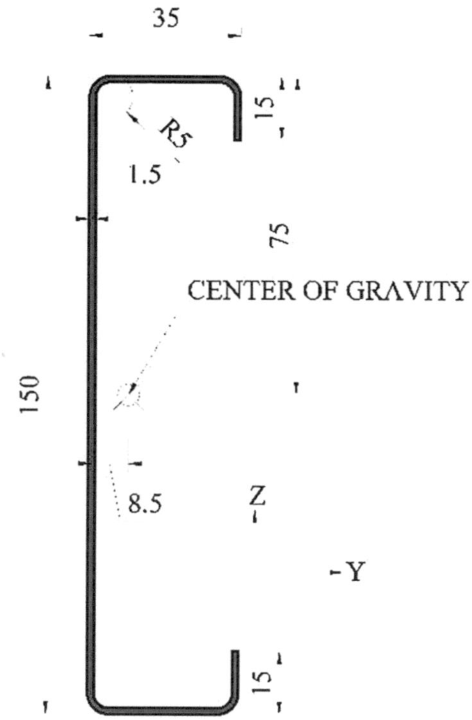

Fig. 25.1 Cross-sectional properties of steel joists

Table 25.1 Geometric and material properties of steel joists

Property	Symbol	Value [units]
Cross-sectional area	A_g	355 [mm^2]
Perimeter	P	476 [mm]
Moment of inertia about Y axis	I_y	1,095,468 [mm^4]
Moment of inertia about Z axis	I_z	55,053 [mm^4]
Plastic section modulus about Y Axis	W_{ply}	17,782.5 [mm^3]
Plastic section modulus about z Axis	W_{plz}	3716.83 [mm^3]
Unit weight	SW	2.79 [kg/m^3]
Modulus of elasticity	E	205,000 [MPa]
Flexural stiffness	EI	4.491×10^{11}
Yield strength	f_y	275 [MPa]

sealing washer diameter was 19 mm. The tensile and shear strength of the screw, as given by the manufacturer, were 13 and 10 kN, respectively. The shear connector is shown in Fig. 25.2.

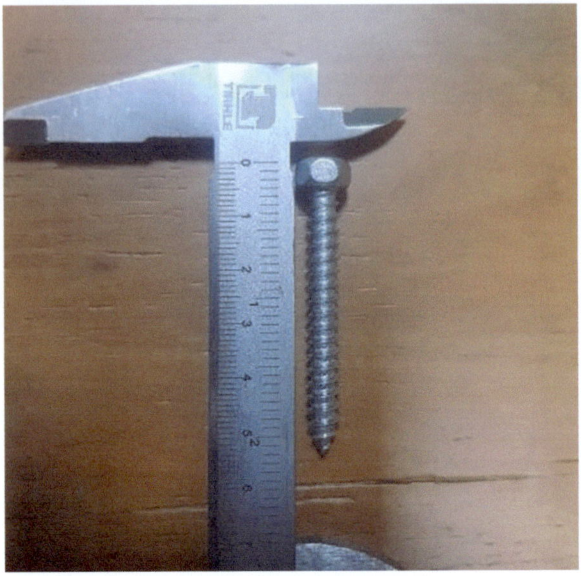

Fig. 25.2 Shear connector

LVL Panels: Laminated veneer lumber (LVL) panels sourced from Eucalyptus grand is and manufactured by Venturer Pte. Ltd. were used for the test. The mechanical properties as declared by the manufacturer, are given in Table 25.2.

Composite Beam Tests: Three four-point loading tests were conducted to assess the strength and stiffness of composite beams with different degrees of shear interaction. The overall test setup is shown in Fig. 25.3. The specimens were simply supported on rollers and loaded at two points using spreader steel beam. The longitudinal spreader beam was IPE 240 with 2.0 m total length. Two transverse beams, UB254 × 146 × 43, were welded to the longitudinal spreader. The center-to-center distance between the transverse beams was 1.4 m. Two solid bars were welded to the

Table 25.2 Material and geometric properties of LVL panels

Property	Symbol	Value [units]
Density	ρ	750 [kg/m^2]
Average bending strength, flatwise, parallel to grain	$f_{m,0,\mathrm{flat}}$	43.24 [N/mm^2]
Average modulus of elasticity, flatwise, parallel to grain	$\mathrm{E}_{0,\mathrm{flat}}$	7450 [N/mm^2]
Average compressive strength parallel to grain	$f_{c,0}$	37 [N/mm^2]
Average tensile strength parallel to grain	$f_{t,0}$	25.95 [N/mm^2]
Width	b	600 [mm]
Thickness	h	70 [mm]
Cross-sectional area	A	42,000 [mm^2]
Moment of inertial about Y axis	I_y	17,150,000 [mm^4]

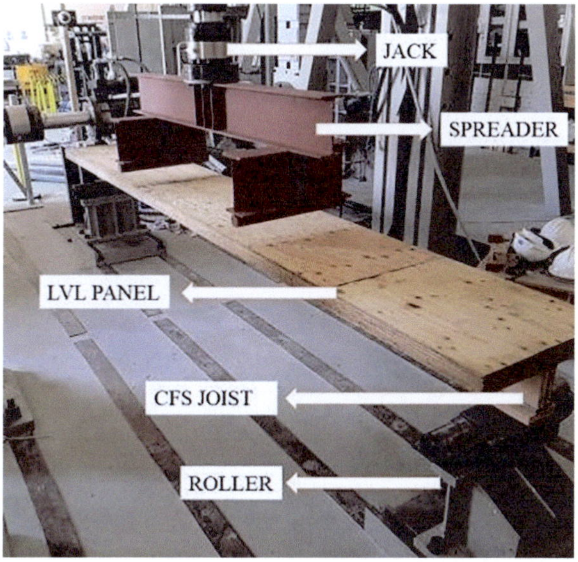

Fig. 25.3 Test setup for the composite beam tests

bottom flange of the transverse beams. Additional steel plates were installed between the specimens and the rounder bars at the load and support points to minimize local indentations. Two variable linear differential transformers (LVDT) were installed at the mid-span point to measure the vertical displacement during the test. The specimens were loaded until failure using a displacement-control system with a 0.02 mm/s displacement rate.

The span for all the composite beams was 4200 mm, with 100 mm extending outside the rollers. In the first beam, the self-tapping screws were spaced at 1200 mm, while for the second and third specimens, the spacing was reduced to 300 and 150 mm, respectively. It should be noted that the steel component of the samples was stiffened with timber infills at the web to trigger the desired failure, and thus avoid any failure relating to local instabilities in the concentrated load areas, Figs. 25.4 and 25.5.

Experimental Results

Modes of Failure: In general, all the specimens failed in a similar pattern. While the LVL panels remained intact after the test, the CFS joists exhibited different degrees of failure under lateral torsional buckling localized near the supports between the fixings. It was also found that the lateral torsional buckling mode depends on the number of shear connectors. Figure 25.6 depicts the failure modes in the CFS joists.

Load–Displacement Curves: The load–displacement curves for each specimen are shown in Fig. 25.7 and Table 25.3. The failure load is the load after which a

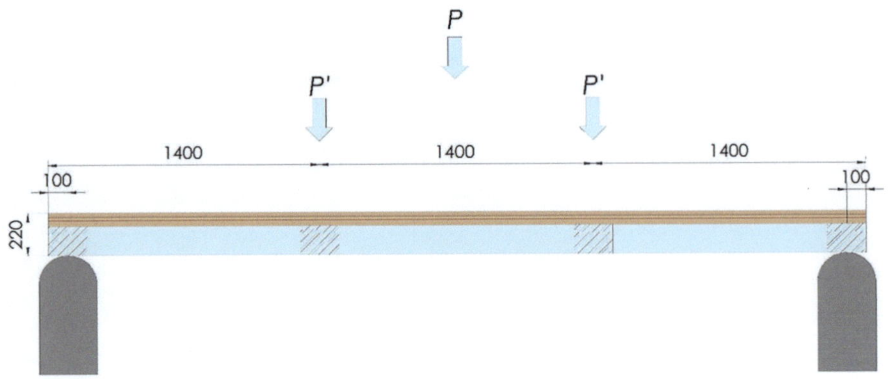

Fig. 25.4 Loading diagram of the composite beams

remarkable and sudden drop in the load occurs. Noticeable twisting in the CFS joists was observed at the failure loads. The specimens continued to resist the load after the point of theoretical failure due to the wooden stiffeners installed at the sides of the webs at the load and support points.

The elastic mid-span deflection for simply supported beam with two concentrated loads can be calculated according to following equation:

$$\delta_{\max} = \frac{23P'L^3}{648EI} \tag{25.1}$$

Denoting P' as one concentrated load, L is the length of the beam between the supports, E and I is the modulus of elasticity and moment of inertia of the composite beams. Therefore, the equivalent flexural stiffness of the composite beam can be calculated from the load–displacement curve up to 25% of $F_{u,\text{exp}}$ (elastic range) according to Eq. (25.1).

Comparison of the Test Results with the Theoretical Bare Steel Resistance: The ultimate moment capacity and the equivalent flexural stiffness of each of the examined specimens compared to the ultimate theoretical elastic moment capacity and the flexural stiffness of the bare steel section are shown in Figs. 25.8 and Fig. 25.9. It can be observed that significant improvements are achieved by mobilizing the composite action. The moment capacity and flexural stiffness of FP1200 were approximately four times higher than bare steel's. Decreasing the spacing of the shear connectors to 300 mm, the moment capacity and flexural stiffness increased by more than five times compared to bare steel. However, shear connector spacing became incrementally less effective when decreased to 150 mm. The moment capacity of FP150 was only 2.6% higher than the moment capacity of FP300, while the flexural stiffness was 20% higher.

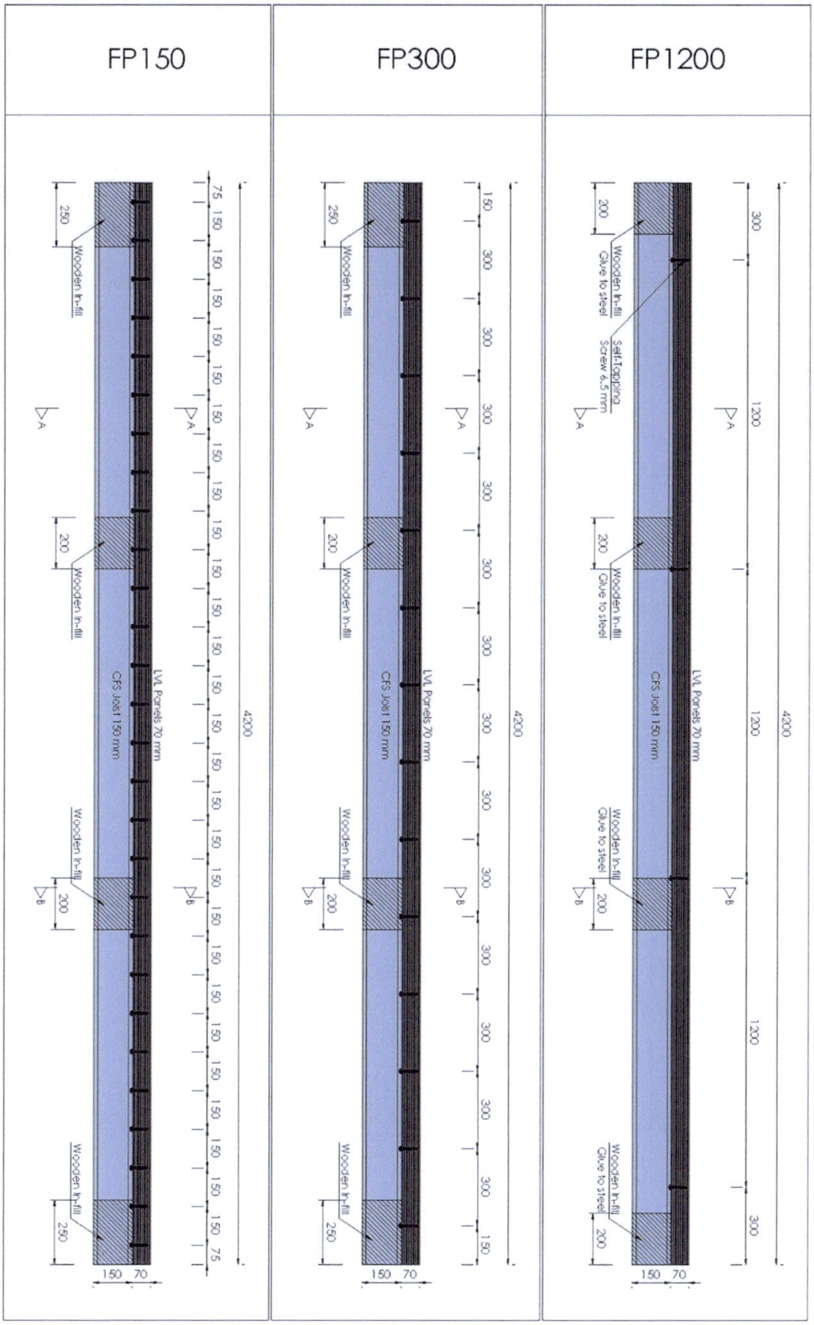

Fig. 25.5 Longitudinal sections of the specimens

Fig. 25.6 Failure modes in the test specimens

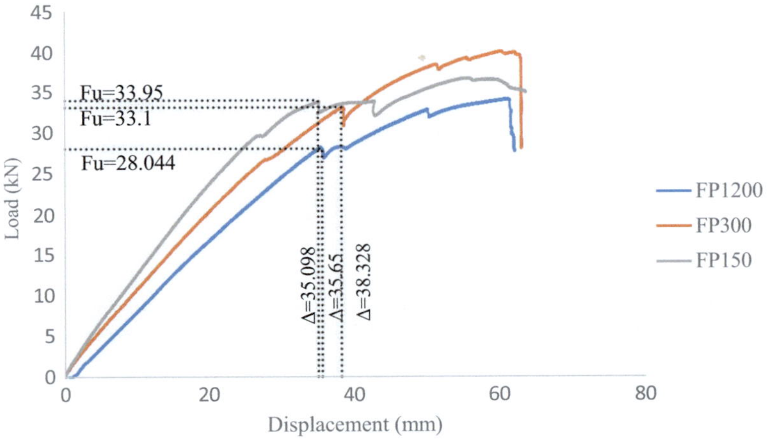

Fig. 25.7 Load–displacement curves

Table 25.3 Equivalent flexural stiffness of the specimens

Specimen	$0.25f_{u,\exp}$ [kN]	P' [kN]	$\delta_{\max}$ [mm]	L [mm]	$EI_{\exp}$ [N mm^2]
FP1200	7.011	3.506	8.957	4000	8.892×10^{11}
FP300	8.275	4.138	7.522	4000	12.5×10^{11}
FP150	8.488	4.244	6.461	4000	14.92×10^{11}

Embodied Carbon Quantification for Two Different Structural Schemes

Two fictional structural systems were created for a three-story residential building in Dubai. The first structure is entirely made of concrete, while in the second system the concrete slab was replaced by the LVL-CFS composite floor panels proposed

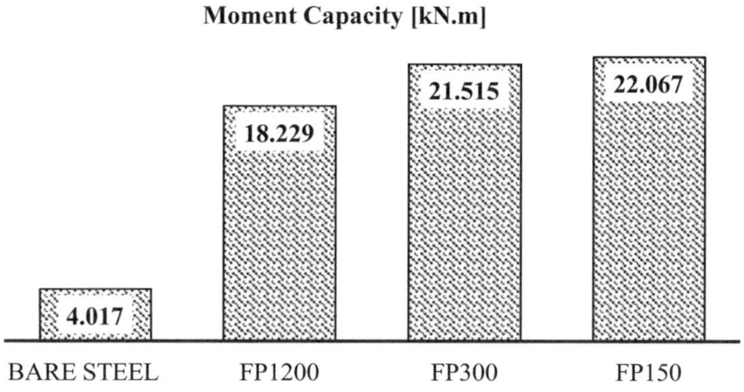

Fig. 25.8 Moment capacity of each specimen compared to the moment capacity of hte bare steel section

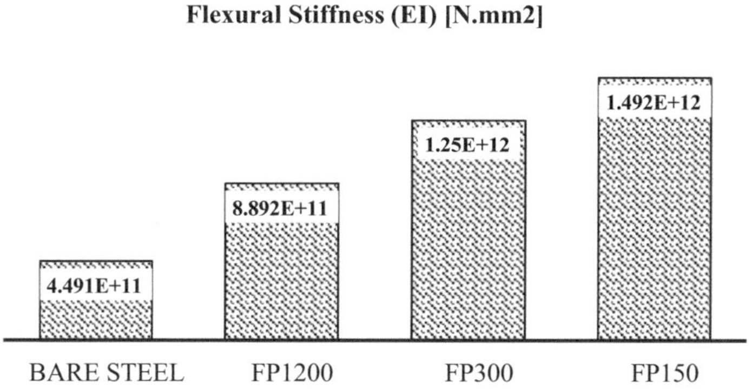

Fig. 25.9 Flexural stiffness of each specimen compared to the flexural stiffness of the bare steel section

previously in this study. The floor height is 3.5 m, and frames consist of three equally spaced bays (4.2 m).

In both systems, concrete was assumed for the single footings and columns. To account for the weight reduction due to composite floors, a reduced size of the footings and columns was selected while stressing the members and the soil beneath by the same magnitude in both systems. The live load in both scenarios was set equal, though the weight of 100 mm screed was not considered the composite floor case.

The concrete beams in the two systems are reinforced with 3T16 rebars top and bottom layers. The ratios of the reinforcement to concrete cross-sectional area in footings and columns are 0.18% and 1%, respectively. The concrete slabs are reinforced with 5T12 per running meter.

The boundary limits are set to production and construction stages (upfront carbon), as the upfront carbon (modules A1–A5) constitutes the majority of the whole life cycle embodied carbon. The calculation of the sequestrated carbon was based on the

Table 25.4 Material quantities

Material	Concrete system		Hybrid system	
	m^3	ton	m^3	ton
Concrete C32/40 (m^3)	165.05	396.13	54.44	130.65
Timber (m^3)	–	–	31.29	23.46
CFS (ton)	–	–	–	2.21
Reinforcement (ton)	–	15.83	–	6.39

assumption that LVL is sourced from a sustainably managed source. The estimated project total cost is three million Euros.

Table 25.4 lists the total quantities of the construction materials.

Carbon Factors

The institution of Structural Engineers (2022) provides guidelines to calculate embodied carbon and suggests embodied carbon factors for common construction materials at each module and stage. The embodied carbon factor represents the total GHG estimated in kgCO2e emitted per one kilogram of the material.

Biogenic Sequestration (Hybrid System)

Given the density of LVL is 750 kg/m^3 and moisture content is 12%, the dry density is $(750/1 + 0.12) = 669.65$ kg/m^3, the biogenic carbon would therefore be: $669.5 \times 0.5 = 334.75$ kg Carbon.

The amount of CO2 sequestrated in 1 m3 of LVL would therefore be $334.75 \times (44/12) = -\ 1227.42$ kgCO2e/m^3. The total amount of CO2 sequestrated in 31.29 m^3 of LVL would therefore be $= -\ 38{,}405.98$ kgCO2e (Fig. 25.10).

Concrete		
A1-A3 (production)	0.109 kgCO2e/kg	Local average*
A4 (transport)	0.005 kgCO2e/kg	ISE, assume local, distance 50 km, Road transport emission, average laden
A5w (waste):	$0.053 \times (0.109 + 0.005 + 0.005 + 0.013) = 0.007$ kgCO2e/kg	ISE
Timber		
A1–A3 (production)	0.390 kgCO2e/kg	ISE
A4 (transport)	$0.005 + 0.020 = 0.025$ kgCO2e/kg	ISE, imported from China, 12,000 km, sea transport

(continued)

(continued)

A5w (waste)	0.111 × (0.39 + 0.005 + 0.005 + 0.13) = 0.059 kgCO2e/kg	ISE
Cold-formed steel		
A1–A3 (production)	2.670 kgCO2e/kg	ISE
A4 (transport)	0.005 kgCO2e/kg	ISE, assume local, distance 50 km, Road transport emission, average laden
A5w (waste)	0.011 × (2.67 + 0.005 + 0.005 + 0.13) = 0.031 kgCO2e/kg	ISE
Rebar		
A1–A3 (production)	1.070 kgCO2e/kg	Local*
A4 (transport)	0.005 kgCO2e/kg	ISE, assume local, distance 50 km, Road transport emission, average laden
A5w (waste)	0.053 × (1.070 + 0.005 + 0.005 + 0.013) = 0.063 kgCO2e/kg	ISE

Results and Discussion

The upfront carbon was calculated for two different structural systems in accordance with the guidelines provided by The Institute of Structural Engineers. The purpose of the comparison was to evaluate the environmental impact of integrating CFS-LVL composite floors with concrete structures. It was found that embodied carbon in hybrid structures is remarkably reduced by 29.6% compared to concrete structures. Having said that, it was concluded that LVL has the second largest contribution to embodied carbon in hybrid structure after concrete (28% of the total embodied carbon). In addition, given that wood is sourced from a sustainably managed resources, implementing LVL in floors could sequestrate nearly 39 tons of GHG emissions. These findings highlight the benefits of relying on timber as floor panels, especially with the enhanced strength and stiffness properties provided by cold-formed steel joists (Figs. 25.11, 25.12 and 25.13).

Conclusion

Three four-point loading tests were carried out on composite floor panels consisting of laminated veneer lumber and cold formed steel to evaluate the degree of shear interaction. It was found that the screw spacing has a significant influence on mobilizing the shear interaction in the flooring system. An increase in the flexural stiffness up to 48% normalized by the non-composite stiffness has been achieved. A profound

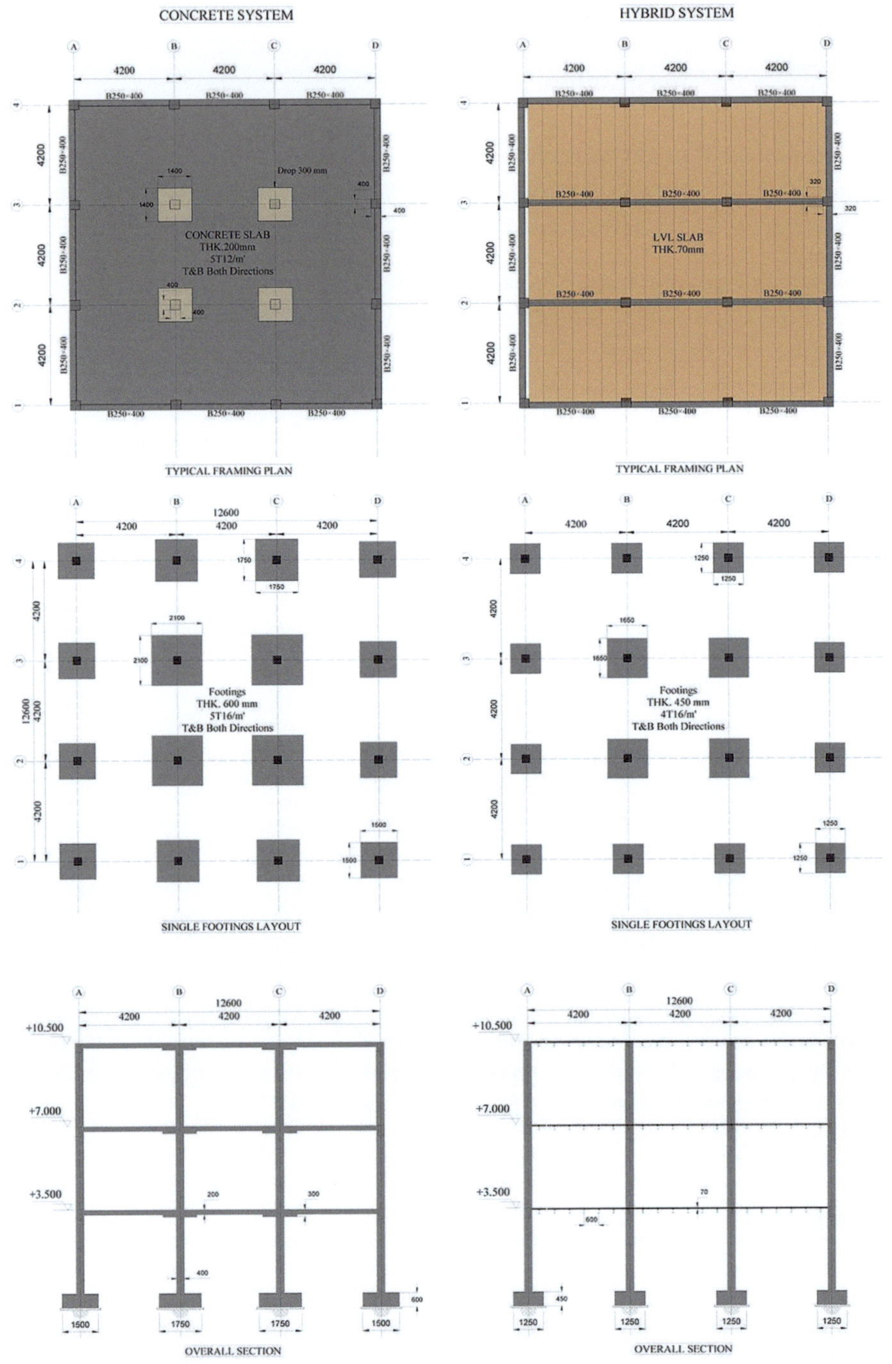

Fig. 25.10 Typical plans and sections

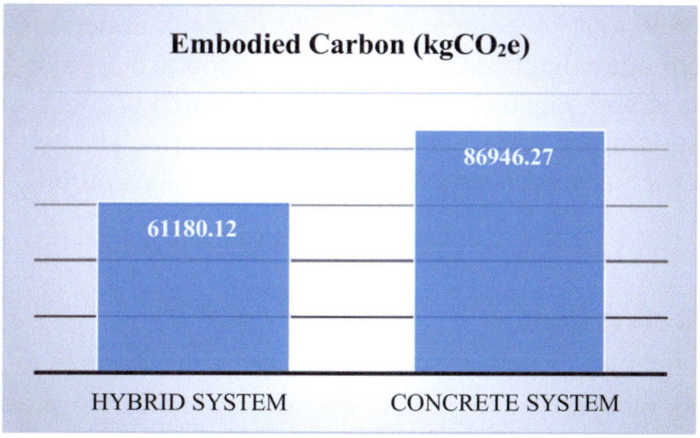

Fig. 25.11 Embodied carbon emitted from each system

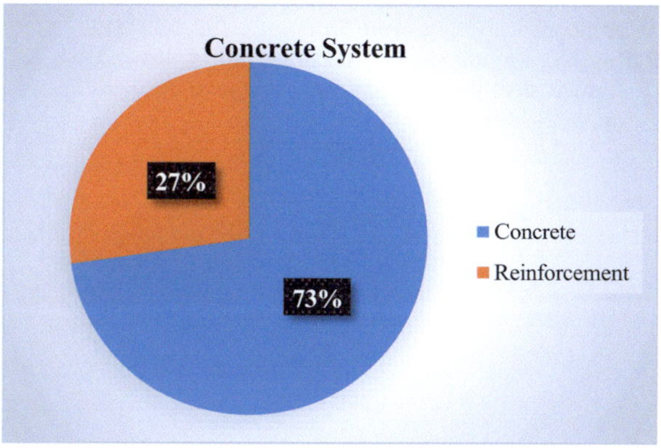

Fig. 25.12 Material contribution to embodied carbon in the concrete structures

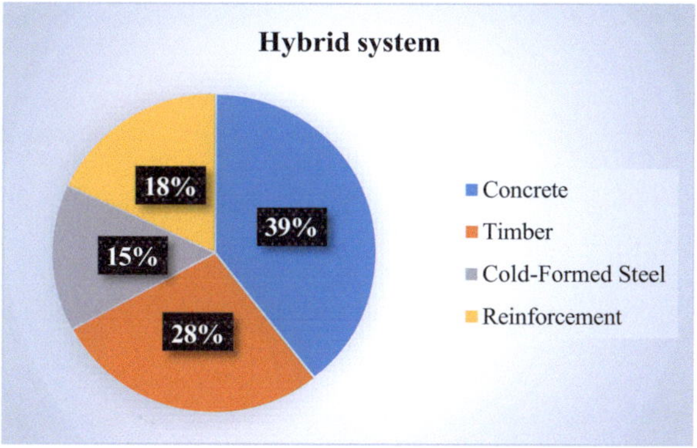

Fig. 25.13 Material contribution to embodied carbon in the hybrid structures

enhancement in the moment capacity of the composite panels was observed with more than five times higher than the moment capacity of bare steel sections.

The results of the environmental impact emphasize the significant role timber in general, and the proposed composite floor system can play in reducing GHG emissions. A comparative study between concrete and hybrid CFS-LVL-concrete structures has been conducted. It was found that embodied carbon in hybrid structures is remarkably reduced by 29.6% compared to concrete structures. More importantly, structural timber is considered carbon sinks. The hybrid structural system was found to sequestrate nearly 38.5 tons of embodied carbon.

References

Kyvelou, P., Gardner, L., & Nethercot, D. A. (2015). Composite action between cold-formed steel beams and wood-based floorboards. *International Journal of Structural Stability and Dynamics, 15*(8), 1540029.

Loss, C., & Davison, B. (2017). Innovative composite steel-timber floors with prefabricated modular components. *Engineering Structures, 132*, 695–713.

Raffoul, S., Moutaftsis, D., Heywood, M., & Rowell, M. (2019). Composite behaviour of cold-formed steel-timber floors. In *ce/papers: Proceedings in civil engineering* (Vol. 3, no. 5–6, pp. 372–380).

Chapter 26
The Role of the Givoni Bioclimatic Chart in Shaping a Sustainable Design

Michela Battaglia

Abstract As the world continues emphasizing sustainable building practices, architects and designers seek new ways to integrate environmental considerations seamlessly into their projects. A useful tool for this purpose is Givoni's Bioclimatic Chart which, based on the climatic zone, informs on the active and passive heating and cooling strategy to achieve the optimal comfort zone for occupants and minimize the use of energy for air conditioning of indoor spaces. The term "bioclimatic" refers to an architectural and urban planning approach that integrates the local climate and natural elements to create energy-efficient and sustainable buildings. It involves using passive design techniques that work in harmony with the surrounding environment, such as optimizing building orientation use of natural ventilation, as well as passive heating and cooling. The use of the diagram in the early stages of design guides architects toward building envelope strategies that can reduce the building's energy consumption by minimizing the use of cooling and heating systems, thereby lowering its ecological footprint while simultaneously enhancing the occupants' well-being. The article describes how to interpret Givoni's Chart and implement its recommendations into the bioclimatic design.

Keywords Bioclimatic building · Sustainability · Sustainable design · Passive design · Bioclimatic strategies

Introduction

The understanding of environmental conditions and their impact on architectural design has been a pivotal aspect of creating sustainable and energy-efficient buildings. Among the tools available for this purpose, Givoni's Bioclimatic Chart stands out as a comprehensive framework for analyzing thermo-hygrometric conditions relative to prevailing outdoor environments. This article delves into the intricacies of navigating

---

M. Battaglia (✉)
AESG, Dubai, UAE
e-mail: michela.battaglia.94@gmail.com

© The Author(s), under exclusive license to Springer Nature Switzerland AG 2026
J. Dsilva (ed.), *Redefining Sustainability*, https://doi.org/10.1007/978-3-031-88376-7_26

the Givoni Bioclimatic Chart, elucidating its components and their implications for architectural design strategies.

Central to the utility of the Givoni Bioclimatic Chart are the 14 color-coded zones overlaying the diagram. Each zone corresponds to specific design strategies tailored to distinct climatic conditions. From comfort zones to strategies for heating, cooling, and solar protection, these zones offer architects a roadmap for optimizing indoor comfort while minimizing energy consumption.

Literature Review

The integration of climatic considerations into architecture has evolved significantly since the early twentieth century. Initially mentioned in the 1930s, the term "bioclimatology" gained prominence with Victor Olgyay's seminal work in the 1950s and 60 s. Olgyay introduced the concept of bioclimatic architecture in his 1953 paper and later expanded it in his 1963 book, Design with Climate. He emphasized an architecture that acts as a filter between humans and the climate, advocating for a harmonious relationship that enhances human comfort by leveraging climatic conditions (Olgyay 1963). Olgyay's bioclimatic chart, which identifies comfort zones considering various weather conditions, laid the groundwork for later scholars. B. Givoni and S. Szokolay further developed these ideas, focusing on comfort zones informed by psychrometric data, reinforcing the importance of climate-adaptive design (Givoni 1992).

This stream of research underscores the enduring relevance of bioclimatic principles in creating sustainable, comfortable, and environmentally integrated architectural solutions. Olgyay's pioneering efforts, followed by Givoni's and Szokolay's enhancements, form the cornerstone of our understanding of bioclimatic architecture, guiding contemporary sustainable design practices.

Methodology

Our study employs the Givoni Bioclimatic Chart to assess climatic data for various climates, including cold, continental, hot dry, and hot humid zones.

Using a focused case study approach, we analyze Dubai's climate (UAE) to demonstrate the chart's application in guiding sustainable design in extreme conditions. This methodology blends qualitative and quantitative analysis to show how the chart informs energy-efficient and comfortable design solutions tailored to specific climatic challenges: from cold climates necessitating passive solar design to hot climates mandating natural ventilation and shading systems, the chart's versatility underscores its relevance across varied environmental settings.

Findings

Givoni's Bioclimatic Chart allows us to analyze thermo-hygrometric conditions according to the prevalent outdoor environment.

Thermo-hygrometric refers to the comprehensive analysis and measurement of temperature relative humidity (%RH), which describes the amount of water vapor in the air.

The graph is divided into two axes where the vertical axis describes the specific humidity g/kg and the horizontal axis describes the dry bulb temperature.

In the diagram, the curved lines represent the relative humidity. It is expressed in percentages, as it corresponds to the ratio of water vapor contained in the air to the water vapor required to saturate one kilogram of dry air at dry bulb temperature.

In addition, 14 color-coded zones are overlaid on the diagram, each corresponding to a design strategy according to specific climatic conditions:

1. Comfort zone;
2. Comfort zone permissible;
3. Heating internal gains;
4. Passive solar heating;
5. Active solar heating;
6. Humidification;
7. Conventional heating;
8. Solar protection;
9. Cooling high thermal mass;
10. Evaporative cooling;
11. Cooling high thermal mass with night cooling;
12. Cooling natural and mechanical ventilation;
13. Air-conditioned;
14. Conventional dry;

By superimposing the characteristic points of a particular climatic zone on the Givoni's bioclimatic diagram, resulting from the intersection of the outside air temperature and the specific air humidity, an overview of climatic conditions is obtained. The concentration of these points on specific areas of the graph allows the choice of design strategies to be applied to the building, to achieve optimal indoor comfort, while reducing the use of energy.

Example

The project is located in Dubai, UAE. The following dry bulb temperatures and humidity data in the following table are extracted from official sources: National Aeronautics and Space Administration (NASA) (National Aeronautics and Space Administration 2022).

Then, selected data for given months are superimposed on the diagram and identified as color-coded points.

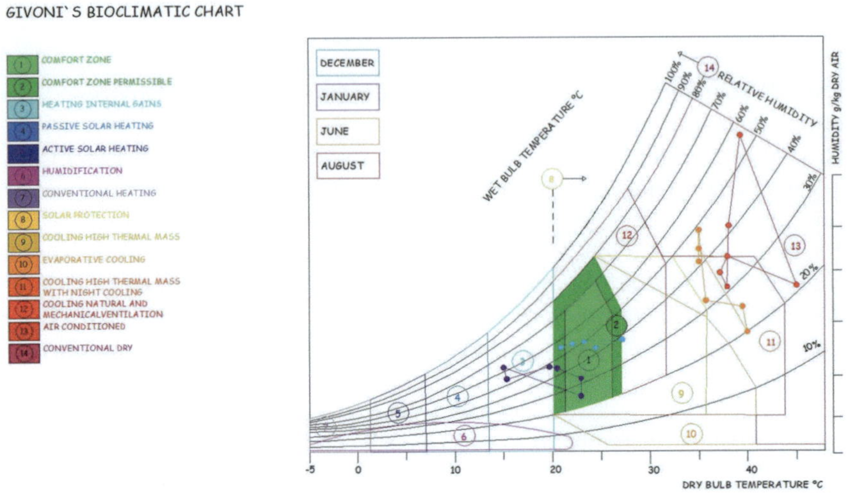

DATA OF TEMPERATURE AND RELATIVE HUMIDITY OF THE YEAR 2022 IN DUBAI,UAE

	0:00		4:00		8:00		12:00		16:00		20:00	
	Temperature (°C)	Humidity (%)	Temperature (°C)	Humidity (%)	Temperature (°C)	Humidity (%)	Temperature (°C)	Humidity (%)	Temperature (°C)	Humidity (%)	Temperature (°C)	Humidity (%)
21-Dec	23	50	22	53	21	56	27	39	27	39	24	44
7-Jan	19	49	16	56	15	63	23	37	23	32	21	46
21-Jun	35	49	35	41	35	44	40	20	39	26	36	32
9-Aug	38	37	37	35	38	31	45	21	39	57	38	42

All the weather data come from NASA's MERRA-2 Modern-Era Retrospective Analysis

Fig. 26.1 Example of Givoni's bioclimatic chart in Dubai, UAE

Next, the recommended strategy is identified by the area that the data points fall into. One can also understand which times of the year and even during the day relate to which strategy to later implement in the design (Fig. 26.1).

In the case of Dubai, there were 4 different periods of the year 2022 analyzed:

December 21, June 21, and the hottest day of the year, (which in 2022 was August 9) and the coolest day of the year (which in 2022 was January 7).

The limits of active and passive cooling and heating strategies are superimposed on the psychrometric diagram.

December

This time of the year is related to the most comfortable conditions of the external environment throughout the day, with most of the time falling into comfort zone 1, with the remaining time falling into permissible comfort zone 2. At the same time, this set of environmental conditions implies the need for solar protection on the glazed area of the building envelope (zone 8).

January

Is considered to be the coldest month of the year. The environmental conditions fall into Comfort zone 1, Permissible Comfort Zone 2, and Zone 3, where the building will heat itself through internal gains.

June/August

During the summer period of June to August, which are considered the warmest months of the year, the environmental conditions fall into zone 11, where the building will be able to cool itself if a strong internal thermal mass is used along with night cooling, and zone 13, considered the extreme climate where the only solution for providing comfort indoor environment is to use air conditioning.

It is important to note, that throughout the year the environmental conditions of Dubai are always in zone 8 solar protection, where solar protection must be used to protect against solar energy entering the building and further heating the interior spaces.

This study shows how the prevalent use of large façade glazing in hot and dry climates, such as Dubai, can challenge sustainability and efficiency despite advances in glass technology. Even with high performance glazing boasting low U-values and high SHGC values, the design approach of heavily glazed façades leads to increased energy demand for cooling due to the strong solar exposure typical of these climates. In contrast, the strategies identified in Givoni's chart advocate a design paradigm that favors more opaque façades combined with high internal thermal mass and the strategic use of shading elements. This approach not only mitigates solar gain but also exploits the properties of thermal mass to stabilize internal temperatures, significantly reducing the reliance on mechanical cooling systems.

Discussion

The implications of our findings extend beyond specific case studies, offering valuable insights for the broader field of sustainable architecture. Theoretically, our research underscores the relevance of Givoni's Bioclimatic Chart as a foundational tool in the architectural design process, advocating for its integration into architectural education and practice. Practically, the study highlights innovative applications of bioclimatic principles, demonstrating their adaptability to contemporary challenges in different climate zone (temperate, hot, cold, humid, etc.) and identifying different strategies, systems, and mechanisms to achieve the optimum comfort zone with minimum energy required. In cold climates, where temperatures hover around 10° centigrade and fall in zone 4 of the table, passive solar design is recommended, optimizing orientation, and adjusting the window/wall ratio to take advantage of solar heat gain during winter, thus reducing dependence on heating systems and minimizing energy consumption.

If we should be in "Zone 5" instead, the table will recommend active solar heating. This means that the thermal conditions inside the building are influenced by an active solar heating system that uses solar energy to actively generate heat.

A solar active heating system is used in zone 5, which may include solar panels, solar thermal collectors, or other technologies that capture solar energy and convert it into heat that can be used for space heating. This solar active heating system can be integrated with an air or fluid distribution system to distribute the heat generated to the different zones of the building (Fig. 26.2).

For buildings situated in hot and sunny climates, natural ventilation, and effective shading systems, such as external louvers, overhangs, and deciduous trees, are recommended to prevent excessive solar heat gains, maintain comfortable indoor temperatures, and reduce the need for cooling (Fig. 26.3).

In arid climates, water sources can be utilized for indoor cooling and passive cooling techniques can be integrated to create pleasant indoor spaces (Fig. 26.4).

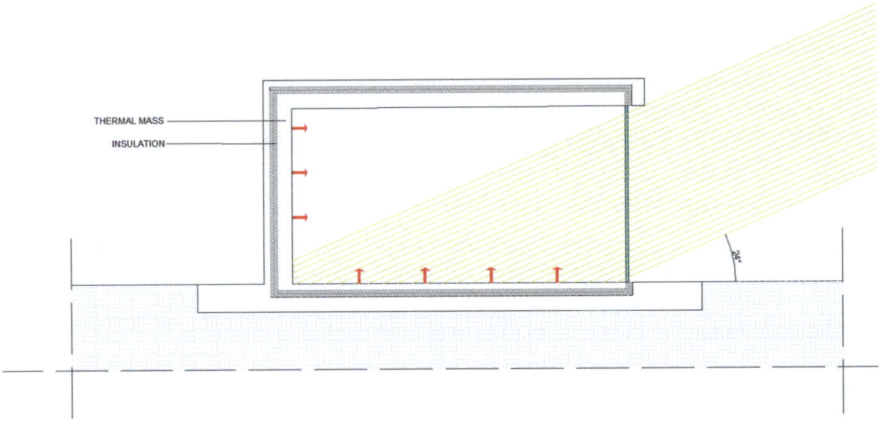

Fig. 26.2 Example in zones 4, 5, 6, and 7 where the orientation of the fenestrations toward direct sunlight brings the benefits of reduced use of energy for heating

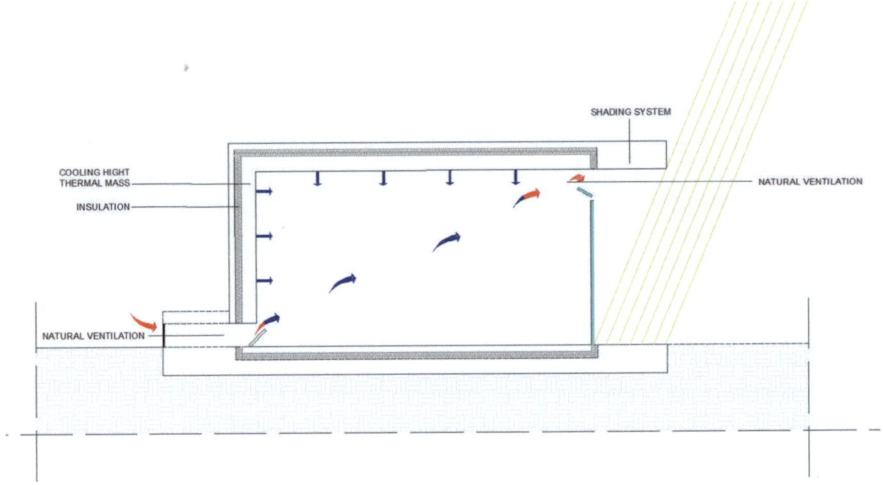

Fig. 26.3 Example in zones 8 and 9, where natural ventilation and shade systems prevent the building's interiors from overheating

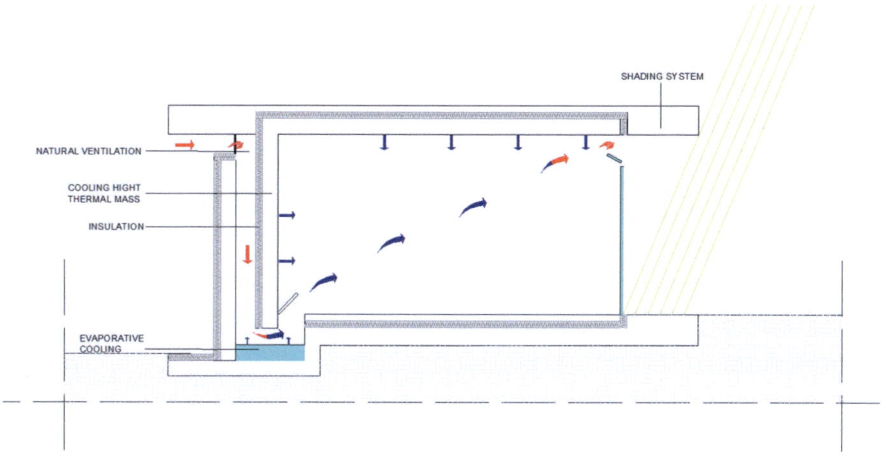

Fig. 26.4 Example in zone 10, where natural ventilation with evaporative cooling and shade systems prevents the building's interiors from overheating

In tropical climates, architects can implement natural ventilation and passive cooling strategies, utilizing shaded outdoor spaces and material moisture-absorbing (Calcium chloride) materials to mitigate high temperatures and humidity, providing a comfortable indoor environment (Fig. 26.5).

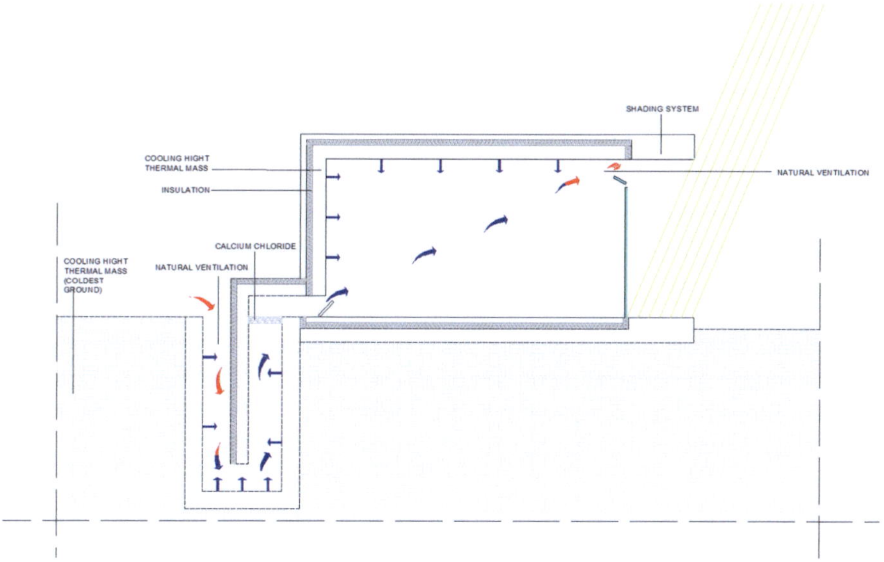

Fig. 26.5 Example in zones 11 and 12, where cooling high thermal mass with cooling ventilation systems prevents the building's interiors from overheating

In temperate climates, Givoni's Bioclimatic Chart encourages architects to incorporate natural ventilation strategies, including operable windows, stack ventilation, and cross-ventilation, promoting airflow and enhancing indoor air quality for improved occupant comfort.

Furthermore, the table highlights the importance of internal thermal mass in façade design. Materials with a high thermal mass, such as concrete or stone, can store heat or cool and then gradually release it over time, contributing to stable indoor temperatures in different climates.

Conclusions

In concluding this article, several key points emerge from the exploration of Givoni's Bioclimatic Chart and its application in architectural design. Firstly, the principles gleaned from our analysis underscore the chart's efficacy in guiding design decisions based on local climatic conditions. By aligning strategies with specific zones on the chart, architects can optimize indoor comfort while minimizing energy consumption, thus fostering sustainable building practices.

However, it is essential to acknowledge certain exceptions, problems, and limitations inherent in this approach. While the Givoni Bioclimatic Chart provides a robust framework for design optimization, its effectiveness may vary depending on factors such as site-specific microclimates or evolving climatic patterns due to climate change. Addressing these nuances requires a nuanced understanding of local contexts and ongoing monitoring of environmental dynamics. Additionally, the implementation of bioclimatic principles may encounter barriers such as cost constraints or regulatory hurdles, necessitating a nuanced approach to implementation.

Despite these challenges, the theoretical and practical implications of leveraging the Givoni Bioclimatic Chart are profound. From a theoretical standpoint, its integration into architectural discourse underscores the importance of adaptive design strategies rooted in climatic responsiveness. Practically, the chart empowers designers to create buildings that not only enhance occupant comfort but also mitigate environmental impact, contributing to broader sustainability goals.

In conclusion, the synthesis of findings underscores the value of the Givoni Bioclimatic Chart as a tool for promoting sustainable and energy-efficient architectural design. By drawing upon the principles elucidated herein, architects can navigate complex climatic dynamics with confidence, yielding buildings that are not only aesthetically pleasing but also environmentally responsible. Moving forward, continued research and refinement of bioclimatic principles will be crucial, ensuring that architectural practice remains adaptive and responsive to evolving environmental challenges.

References

Givoni, B. (1992). Comfort, climate analysis, and building design guidelines. Retrieved from https://www.sciencedirect.com/science/article/abs/pii/037877889290047K

National Aeronautics and Space Administration. (2022). MERRA-2. NASA. https://gmao.gsfc.nasa.gov/reanalysis/MERRA-2/

Olgyay, V. (1963). Design with Climate.

Chapter 27
Assessing Architects Perspective on Using Sustainable Materials in the UAE Construction Sector

Aditya Pramodu Pillai and Anjali Arun Nair

Abstract Rapid urbanization in Gulf nations has spurred infrastructure growth, necessitating sustainable practices to mitigate environmental impacts. The main aim of this study is to explore the adoption of sustainable materials in the UAE construction sector through a qualitative assessment of 16 architects in the UAE. A detailed analysis reveals the emergence of four themes which highlight the current sustainable practices emphasizing resource efficiency and environmental consciousness, challenges persisting in material selection, regulatory complexities and limited local availability. And benefits including reduced carbon emissions and operational expenses, enhanced building longevity and occupant well-being, and finally sustainability drivers such as regulatory frameworks and growing corporate and consumer awareness. The study underscores the importance of sustainable materials in achieving UAE's sustainability goals, emphasizing the need for tailored strategies amidst regional and economic dynamics.

Keywords Sustainable materials · Environmental impacts · Sustainability drivers · Qualitative

Introduction

Rapid urbanization and the increasing urban population have significantly accelerated infrastructure development worldwide, making the construction industry one of the fastest-growing sectors globally, particularly in Gulf nations (Govindan et al., 2016). However, this rapid expansion has introduced significant environmental challenges, including increased energy consumption, resource depletion, and carbon emissions (Radhi, 2010). Sustainability, in the context of construction, refers to the practice of

A. P. Pillai (✉)
Vellore Institute of Technology, Vellore, India
e-mail: adityapillai2003@gmail.com

A. A. Nair
Arif and Bintoak Consulting Architects and Engineers, Dubai, UAE

© The Author(s), under exclusive license to Springer Nature Switzerland AG 2026
J. Dsilva (ed.), *Redefining Sustainability*, https://doi.org/10.1007/978-3-031-88376-7_27

creating buildings and infrastructure that meet present needs without compromising the ability of future generations to meet their own needs (Ahmed & El-Sayegh, 2022). This involves the use of sustainable materials and construction practices that minimize environmental impact and promote resource efficiency buildings (Danso, 2018). In the UAE, most construction projects prioritize aesthetics and cost when selecting materials, often neglecting energy and environmental performance. Consequently, 70% of the country's annual electricity consumption is attributed to building systems, leading to a dramatic rise in energy consumption and making the UAE one of the highest per capita energy consumers globally (Kazim, 2007). As the UAE strives to balance developmental goals with environmental sustainability, the construction industry plays a crucial role. Central to this effort is the adoption of sustainable materials, which can mitigate environmental impacts and enhance the long- term viability of buildings (Ali & Al Nsairat, 2009).

The UAE's harsh desert climate and limited natural resources further necessitate the integration of sustainable materials to address environmental challenges and promote resource conservation (Jaillon & Poon, 2014). To promote sustainability in construction, the UAE government has launched several initiatives, including the Estidama Pearl Rating System and Dubai's Green Building Regulations, aimed at encouraging the use of environmentally friendly materials and practices (Aboul-Naga & Elsheshtawy, 2001). Early research (Wastiels & Wouters, 2008) advocate that architects have the responsibility to select the right materials for the environment. Despite these efforts, there is limited research focusing on the perspectives of architects in the UAE regarding sustainable materials. Therefore, the objectives of this research are fourfold in nature. Firstly, to identify the current sustainable practices in terms of materials employed in the UAE construction sector. Secondly to determine the challenges faced by architects in obtaining sustainable materials in construction in the UAE. Thirdly, to assess the motivations behind using sustainable materials in the UAE construction industry and finally evaluate the benefits of using sustainable materials in construction for consumers and the region from an architect's perspective. Understanding these perspectives is crucial for identifying barriers and facilitators to the widespread adoption of sustainable practices in the region. Also, to date, there are no studies examining the role of architects and their views on the use of sustainable materials in the UAE construction sector.

Theoretical and Empirical Review

Innovation diffusion theory aims to comprehend the mechanisms, reasons, and speed at which new ideas and technologies disseminate within a social system (Rogers, 1962). The theory of innovation diffusion (Rogers, 1962) serves as a theoretical underpinning in assessing architects' perception towards sustainable materials (Markström et al., 2018). The theory highlights that the adoption rate depends on how individuals perceive the innovation regarding parameters like utility, compatibility, and complexity. It highlights the roles of individual perceptions (architects in this paper), external factors, environmental concerns and the characteristics of the innovation itself in shaping the adoption process. The theory of planned behaviour (TPB) (Ajzen, 1985) is another crucial theory employed in exploring environmental psychology and green behaviour (Boiral et al., 2015), precisely considering the intention and extent of subjective control that individuals, such as key personnel (architects in our paper), have in their decision-making power. As the TPB postulates that individual intention is the key determinant of any behaviour, here we contend that architect's intention to engage in sustainable practices and relentless endeavours to enforce the performance of sustainable actions results in sustainable behaviour. Additionally, architects have authority and control over their actions, ability, and power to ask for management support apart from the self-will and choice to engage in a sustainable behaviour (see Pillai et al. 2024).

Olivetti and Cullen (2018) review six types of construction materials made from agro-waste: brick/masonry elements, green concrete, insulation materials, reinforcement materials, particleboards, and bio-based plastics. These materials were chosen for their popularity and extensive use in contemporary construction practices. The findings indicate that agro-waste-based construction materials are effective and meet established building standards. This suggests that agro-waste materials have the potential to replace conventional construction materials, promoting economic, environmental, and social sustainability in the long term (see also Schneiderman & Hillmyer, 2017; Pollock, 2010). Additionally, Nilimaa (2023) reviews the current trends and opportunities in sustainable concrete construction, focusing on the adoption of eco-friendly practices to mitigate the industry's environmental impact. The paper explores sustainable materials such as green concrete, supplementary cementitious materials, permeable concrete, cool concrete, and the use of local materials. Additionally, it discusses innovative technologies including self-healing concrete, 3D-printed concrete, photocatalytic concrete, electrified machinery, and carbon capture, utilization, and storage (CCUS) principles, highlighting their potential to enhance construction sustainability. The findings reveal the challenges faced in implementation due to economic and social barriers.

The paper emphasizes the roles of governments, industry, and academia, stressing the need for interdisciplinary collaboration and research (see also Adesina, 2020; Aghimien et al., 2019; Akeed et al., 2022). More recently, Ilgin et al. (2024) explored Finnish architects' perspectives on using timber as a structural material in multistorey residential buildings (exceeding two storeys). The quantitative study gathered responses from 147 architects to infer their perceptions, attitudes and interest in using wood for multi-storey residential buildings. The findings revealed that wood is a lightweight, local, and eco-friendly material but was seen as more expensive and requiring more intricate engineering (see also Mallo & Espinoza, 2015; Markstrom et al. 2018). Generally, the architects had a positive attitude towards using wood in low-rise residential buildings; their views on tall timber buildings were largely negative. Meanwhile, Umoh et al. (2024) conducted a literature review that emphasized green architecture and energy efficiency which are essential in modern construction due to increasing concerns about environmental degradation and energy use. This paper reviews innovative green architecture design and construction techniques to enhance energy efficiency by focusing on sustainable design principles to reduce buildings' ecological footprints, utilizing strategies like passive design, green roofs, and efficient insulation to intelligently harness natural resources. Moreover, innovative construction materials, such as recycled and locally sourced options, and advanced technologies like bio-concrete and self-healing materials were found significant for their structural resilience and eco-friendly properties.

Prior research has extensively studied sustainable construction materials; however evident gaps remain in terms of cultural and regional specificity as their adoption may vary depending on the regions. A severe dearth of studies exists in the UAE economic setting where sustainable research is vital as it aligns with UAE initiatives such as the UAE Vision 2021 and UAE Green Agenda 2030, which emphasizes sustainable development and innovation. Furthermore, this paper endorses national goals and is essential for projects like Masdar City and smart cities, requiring advanced materials for energy-efficient, resilient, and eco-friendly urban environments.

Data and Methodology

Data for the paper has been derived from 16 architects residing in the UAE using purposive sampling technique. Fugard and Potts (2015) posit that qualitative studies require a minimum sample size of 12 to reach data saturation; therefore, a sample size of 16 is deemed sufficient for the study. Architects are chosen as the respondents as they are vital decision makers in material selections during construction (Conroy et al., 2018) and have the potential to enforce sustainable practices in the field (O'Connor et al., 2004) thus transitioning to low carbon buildings (Fischer & Guy, 2009) (as cited in Ilgin et al. 2024, p. 3). A qualitative technique has been adopted to analyse the data collected by conducting structured interviews between December 2023 and February 2024. The respondents were ensured identity anonymity and were assured confidentiality in responses provided. They were also provided the liberty to withdraw from research at any point in time. The paper employs Braun and Clarke (2006) thematic analysis as a robust tool for qualitative data analysis and is particularly useful when a research project aims to discover themes and concepts embedded throughout qualitative data.

Results and Findings

The findings highlight that, of the 16 architects interviewed, 50% were females and 50% males. The majority of the respondents had more than 15 years of working experience in the UAE construction field with the others having more than 10 but less than 15 years' experience. Regarding their profile, the majority of them were project architects and senior architects, with a small percentage of junior architects, all with expertise in architecture and interior design. The codes generated from responses were then grouped together to form four themes as follows:

Theme 1: Sustainable Construction Practices /Materials

Findings revealed that current sustainable building principles focuses on developing structures that are resource and environmentally conscious throughout the duration of their lives. Also, using renewable energy sources, energy-efficient systems, and environmentally friendly materials are important practices followed in the UAE. In addition to promoting better living conditions through enhanced indoor air quality and natural lighting, sustainable building places a strong emphasis on waste reduction, material recycling, and resource conservation. For example, R1 opinioned that "the installation of solar panels and smart home automation systems aims to optimize energy consumption. Green building certifications, like BREEAM and LEED, direct these initiatives by establishing sustainable benchmarks. Regarding sustainable materials, sustainable timber sourcing, low-emission concrete, and recycled steel were just a few of the environmentally friendly elements that the UAE is using frequently in its construction projects. Furthermore, R7 mentioned that "the use of sustainable materials is encouraged by green certifications like LEED and Estidama, which incentivize builders to embrace more ecologically friendly techniques".

Theme 2: Challenges in Promoting Sustainable Buildings

The findings revealed that despite long-term benefits, one significant obstacle is the initial cost of green buildings, which might discourage investment. Here is where it is important to demonstrate the economic advantages of sustainable practices and to raise awareness of them. For example, R4 reported that "obtaining sustainable materials for construction in the UAE presents several challenges, including limited local availability, which often necessitates importing materials at higher costs and with increased carbon footprints". Results also indicated that access to sustainable materials and technology was limited and the imperativeness of government and regulatory frameworks to offer to assist green construction projects. Another respondent R13 reasoned that "navigating the regulatory environment can be complex. The UAE has specific sustainability standards and certifications, such as Estidama and the Dubai Green Building Regulations. Ensuring materials meet these stringent standards can be a challenge". Limited training, social and economic factors and education and stakeholder discord were other prominent challenges (see Nilimaa, 2023 for similar findings).

Theme 3: Benefits and Drivers of Using Sustainable Materials in Construction

Findings evince that the preservation of natural resources and alleviation of detrimental environmental effects due to a decrease in waste and carbon emissions emerge as the main benefits from employing sustainable materials in construction (see Umoh et al., 2024 for similar results). Over the course of their lifetime, these buildings gradually lowered operation costs and increased energy efficiency. For example, R15 remarked that "additionally, the long-term cost benefits of sustainable materials, such as reduced energy consumption and maintenance costs, make them economically attractive despite higher initial investments". R11 mentioned that "improved indoor air quality and natural lighting lead to healthier spaces that improve productivity and occupant well-being". Regarding the drivers, respondents expressed appreciation towards government incentives, surging regulatory pressure, together with growing consumer and corporate knowledge of environmental problems, as the drivers behind these practices. For example, R5 noted that "Some companies are driven by regulatory policies and incentives like tax benefits, while others see the potential for cost savings over the building's lifetime" and R8 remarked that "following regulatory frameworks (Estidama Pearl Rating System, LEED certification)", sustainable building approaches promote a more sustainable future by putting lifespan, flexibility, and a small ecological imprint first. This benefits both the earth and its people.

Theme 4: Benefits of Sustainable Materials in Construction

Research findings denote several benefits accruing from using sustainable building materials in construction, such as reduced environmental impacts and enhanced efficiency. For example, R3 reasoned that "it not only enhances cost savings and improves living conditions for individuals but also makes a positive impact on the environmental, social, and economic aspects of the region. Low-emission concrete, bamboo, and recycled steel all reduce carbon footprints and preserve natural resources. They frequently increase energy efficiency, which lowers utility bills and encourages long-term savings". Also, R15 added that "for the region, it reduces the environmental footprint, promotes resource conservation, and aligns with the UAE's vision for sustainable development and a green economy". Overall, the results suggested that sustainable materials improve indoor air quality and promote better living and working conditions (see Umoh et al., 2024 for similar findings). Their application also enhances a building's market value and appeal by supporting green certifications such as LEED.